STEM LITERACIES IN MAKERSPACES

Providing an original framework for the study of makerspaces in a literacy context, this book bridges the scholarship of literacy studies and STEM and offers a window into the practices that makers learn and interact with. Tucker-Raymond and Gravel define and illustrate five key STEM literacies—identifying, organizing, and integrating information; creating and traversing representations; communicating with others for help and feedback during making; documenting processes; and communicating finished products—and demonstrate how these literacies intersect with making communities. Through careful observation and analysis of multiple case studies, the authors highlight the impact of research and practice to support teaching and making in a variety of environments. Using a nuanced, engaging framework, they examine the necessary skills required to develop and foster makerspaces in formal and informal contexts for all students. Grounded in cutting-edge research, this volume paves the way for future study on supporting making and literacies in STEM.

Eli Tucker-Raymond is a senior research scientist with the Chèche Konnen Center at TERC, a non-profit STEM education research and design organization in Cambridge, MA.

Brian E. Gravel is an assistant professor of STEM Education and Director of Elementary Education in the Department of Education at Tufts University, USA.

STEM LITERACIES IN MAKERSPACES

Implications for Learning, Teaching, and Research

Eli Tucker-Raymond
Brian E. Gravel

Routledge
Taylor & Francis Group
NEW YORK AND LONDON

First published 2019
by Routledge
52 Vanderbilt Avenue, New York, NY 10017

and by Routledge
2 Park Square, Milton Park, Abingdon, Oxon, OX14 4RN

Routledge is an imprint of the Taylor & Francis Group, an informa business

Library of Congress Cataloging-in-Publication Data
Names: Tucker-Raymond, Eli, author. | Gravel, Brian E., author.
Title: STEM literacies in makerspaces : implications for learning, teaching, and research / Eli Tucker-Raymond, Brian E. Gravel.
Description: New York, NY : Routledge, 2019. | Includes bibliographical references.
Identifiers: LCCN 2018050624 | ISBN 9780815367734 (hbk) | ISBN 9780815367741 (pbk) | ISBN 9781351256728 (ebook)
Subjects: LCSH: Maker movement in education—Case studies. | Makerspaces—Case studies. | Science—Study and teaching—Case studies. | Interdisciplinary approach in education—Case studies.
Classification: LCC LB1029.M35 T84 2019 | DDC 507.1—dc23
LC record available at https://lccn.loc.gov/2018050624

ISBN: 978-0-8153-6773-4 (hbk)
ISBN: 978-0-8153-6774-1 (pbk)
ISBN: 978-1-351-25672-8 (ebk)

Typeset in Bembo
by Apex CoVantage, LLC

This book is dedicated to Susan Klimczak and Ashley Freeman, makers and youth advocates for life. And to Nathaniel, Owen, Celia, and Molly.

CONTENTS

FIGURES

TABLES

ACKNOWLEDGMENTS

This project is a celebration of the incredible work that makers, artists, engineers, craftspeople, and educators do in making spaces around the Greater Boston area. We were humbled by the creativity, ingenuity, improvisation, and commitment that we encountered throughout our journey. The ways your communities amplify talent and expertise as you gather together to make things is truly impressive. And just as it takes communities of makers to build these making spaces, it took a community of collaborators to conduct the research presented in this book. We would first like to acknowledge the participants in our study, we appreciate your generosity, kindness, and willingness to be followed by someone, or someones, with a notebook and a camera. Without your commitment to the project, this book would not have been possible.

We must also thank the STEMLiMS research team for enduring the endless schedule of fieldwork, rounds of coding, and obscure references in research team meetings. You each contributed invaluable insights and intellectual contributions to the overall project. This book is a reflection of our work together and our distributed expertise, and we thank you, specifically: Ada Ren, Aditi Wagh, Kaitlin Kohberger, Kyle Browne, Brianna Jefferson, and Vitoria Fonseca. Kyle Browne drew the illustration on the front cover.

To the Inmates at the Asylum, you opened our eyes to new ways that people can create communities, by exploding barriers and boundaries in art, design, and engineering. Thank you to Molly Rubenstein, Derek Seabury, Jess Muise, and Chris Templeman for your support and conversation. Thanks to Calvin Domenico, Aaron Falk, and Edison Geiswein for talking with us and letting us observe their design and fabrication expertise first hand.

To our partners at South End Technology Center and the Menino Center, you welcomed us into your communities to witness the incredible work the

L2TT2L youth, mentors, and organizers (that's you!) do; we are forever grateful for our time with you all, and hope our partnerships continue for years to come. A special thank you to Susan Klimczak, Mel King, Donna Parker, Amon Millner, and Ed Baffi.

Our friends and partners at Nedlam's Workshop, you shared a set of keys and you shared your trust in partnering with us to explore exciting new ground bringing a making space to a public high school. It would not have been possible without your courageous spirit and commitment to the students and your school community. We have learned a lot at Nedlam's, and we are excited to see how the work together continues to grow. A special thanks to Ashley Freeman, Shereen Escovitz, Miki Vizner, Trang Do, Arturo Muñoz, R. Ben Shapiro, and Dana Brown,

We would also like to thank our evaluators, The Program Evaluation and Research Group at Endicott, specifically Gretchen Porter, Debra Smith, and Judah LeBlang, for serving as true critical friends throughout the project. Your work allowed us to see ourselves, where we were succeeding and where we needed to improve. Your collective insights supported our efforts to stay focused and committed to the long-term goals for this project.

To our advisors and colleagues at TERC, Tufts, and beyond: your wisdom, insights, curiosity, and critical eyes pushed and shaped the work in this book. Thanks to Christopher Wright, Andee Rubin, David Hammer, Paula Hooper, Ann Rosebery, Beth Warren, Al DeSena, Vanessa Svihla, Nathan Holbert, Erica Halverson, and Victor Lee.

Eli: A big thank you to Brian for taking my ask for a job and turning it into a five year (and counting) collaboration, and for introducing me to making communities locally and around the country. This work has sparked a new and vibrant trajectory. Thank you to Ellen for helping to remind me what is important and what we are working for—Justice. And thank you to Nathaniel for your support, and for asking daily how the book was going and if it was finished.

Brian: A special thanks to my collaborator, Eli, and everyone at TERC for always making me, and our work, feel at home, valued, and important to the ERC Community. And to Jenna, my champion and thought-partner-in-dinner-making. I can always count on you to indulge my rants, check my assumptions, and help me sharpen my thinking. I feel lucky to have you as a partner in life, and grateful for all that you have done to help this project feel like a success.

This material is based in part upon work supported by the National Science Foundation under Grant Numbers DRL-1422532 and IIS-1450985. Any opinions, findings, and conclusions or recommendations expressed in this material are those of the author(s) and do not necessarily reflect the views of the National Science Foundation.

INTRODUCTION

> Samuel, a Haitian American high school student classified as ELL, is working in an after school making space at his high school. He is attempting to make a safe box for cell phones so that he and his peers can charge their phones during school without them being stolen. First, he talks to his peers about what they think would be a good design. Online, he then researches materials and designs. He looks at pictures and schematics, and reads written language. As he does, he jots notes and communicates with a variety of people. He emails a locksmith for advice. Then he sketches and diagrams a design for a lock. On his diagram are technical words associated with mechanical engineering: drive cam, wheel fly, and spindle, to name a few (see Figure 0.1). Samuel gives an oral presentation of his prototype. When his school principal comes to visit and is enthusiastic about his prototype, Samuel tells him, "See, I'm smarter than you think I am." Samuel's practices, literacies included, allowed him to reposition his academic identity as a successful engineer while creating a product that benefited the people around him.

That is where the real life events end. If we imagine him continuing, Samuel might learn how to use the laser cutter or the 3D printer to make a working prototype. To do so, he would need to learn a software program that would communicate his drawings to the manufacturing tools. A particular software is made up of a whole system of representations. If he were at Artisan's Asylum, one of the exclusively adult making spaces we studied, he might learn professional Computer Aided Drawing (CAD) and Manufacturing (CAM) tools to design his lock to very specific measurements. Most certainly he would interact and communicate with the members of Artisan's Asylum's local making community, potentially widening his social and professional network, and learning from their expertise in any number of areas. He might have to negotiate a business agreement with one of those members to assist in

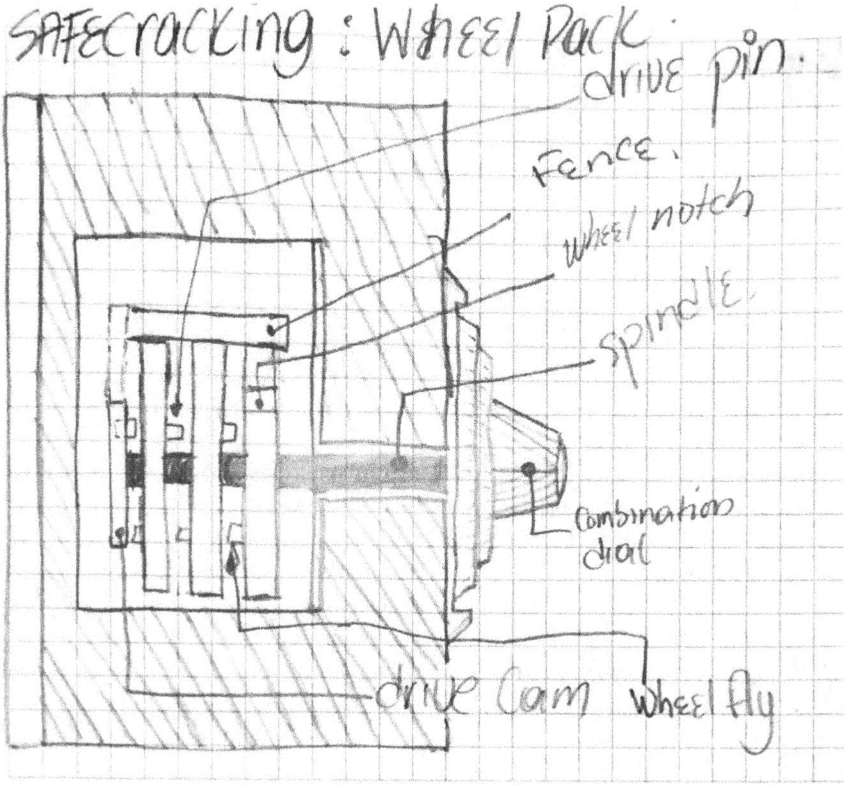

FIGURE 0.1 Samuel's design sketch of his cell phone lock box.

the manufacturing of his idea. Although Samuel did not go on to manufacture a complete working model of his cell phone lock box, what we describe is precisely what experienced adult makers do with representations as they make. They learn new representational systems, translate ideas into different representational forms, communicate with others about their ideas and to gather information, and share their ideas and work with others. They are doing literacy.

Overall Approach to the Study

This book is about learning at the intersection of making spaces; literacies; and science, technology, engineering, and mathematics (STEM). The overall study is grounded in ethnographic and design-research approaches (Brown, 1992; Collins, Joseph, & Bielaczyc, 2004). It was structured as a three-year exploration of making in different contexts to weave together observations from informal and formal spaces, from work with adults and youth, and from formal classroom spaces. The three-year sequence was structured as follows:

a) Interviews with experienced makers from a diverse range of making domains—e.g., artists, engineers, computer scientists, crafters, jewelers, and entrepreneurs; b) observations at two adult making spaces—one a community space, and the other operating as a design and engineering firm; c) observations in a youth making program housed at a community technology center; and d) observations and designed interventions with teachers and students at a high school.

Over three years, we worked closely with four adult and youth making spaces. We immersed ourselves in these communities, observing, interviewing, and working with community members for more than six months at each site. We built from what we learned to design and study teacher professional development activities and a maker-based course taught at a comprehensive urban high school. We spent hundreds of hours conducting interviews and observations in the four different spaces.

The design of the study was intended to first allow us to map the space of literacies in making, beginning with experienced makers that represented different approaches and perspectives. From that, we explored how youth engage literacies in making, not to determine whether their practices mirrored those of the seasoned makers, but rather to complicate the descriptions of practice we established from observations of experienced adults. Our driving goal was to develop the STEM Literacies in Making Spaces framework, and we did this using empirical evidence gathered across different configurations of making.

Another key component of the study was to build from these ethnographic discoveries to construct designs that would allow us to further explore STEM literacies, or what we also call STEM literacy practices. Taking a design research approach, these interventions were conjectures—carefully constructed prompts, protocols, and arrangements that both deepened our inquiry and contributed to theory building within the framework. These interventions are described at points in the book, some include: an expertise mapping exercise with youth, professional development workshops for teachers, and participation and activity structures to support students in developing particular literacies in classroom contexts. Some interventions were only enacted a single time, others were iterated to refine both the approach as well as our understandings of the ways literacies operate in making, and their relationships to STEM disciplines.

Thus, the study began by observing making in practice, with individuals from different locations, and we then used those observations to design ways of further exploring the role of literacies in making. Our methodology hinges on the partners in this work—the individuals and the making spaces that agreed to be part of this study. We are deeply indebted to their willingness to take part in our study and their curiosity in our questions and methods, helping us improve the work every step of the way.

Research Questions

The three-year project was guided by three overarching research questions, each relating to lines of inquiry tailored for the particular sites of study in each year. The questions are these:

- What STEM literacy practices do established makers and adult maker groups identify as important to their work? What STEM literacy practices do they use while participating in different making space activities and how do they engage in them? How are they learned?
- What STEM literacy practices do beginning and developing makers use in making spaces and how do they learn and engage in them?
- How can mentors and facilitators in making spaces support beginning makers' appropriation of literacy practices?

This book is our attempt to distill, analyze, and describe what we saw, both through observations and through our designed interventions. It is a contribution to theories about learning through making and to recognizing and supporting literacy practices in making spaces. It suggests future research directions principles for learning designs at the intersections of making and literacies in STEM.

Making

Making is many different things. People use a vast array of materials to bring their ideas to life, from the oils, essences, soaps, and chemical compounds in toiletries, to the steel bronze and copper of large-scale sculptures. From the acrylic in puzzles lined with copper contacts that when finished tell a story on a computer screen, to the microcontrollers at the core of interactive public art installations, and the fabrics and electric threads of e-textiles. For all of those people, facility with representations—be they written, images, or oral—is key to undertaking successful projects. We call this facility *literacy*. As Samuel showed, literacies are central to making.

We define *making* as engagement in the personal construction of new objects or the hacking/repurposing of existing ones, often, but not necessarily, employing digital and analog technologies together (Halverson & Sheridan, 2014). The argument for making across educational contexts is built on sociocognitive principles of learning. First, people actively make meaning by internalizing, and thus transforming, their social worlds, including the meanings, values, and attitudes available to them (Vygotsky, 1978). Second, people learn best when they are able to transform again, or make external, what they have internalized, and then share it with others in the form of a public artifact (Papert, 1980). Making spaces embody these principles as a social practice, less focused on individuals but rather accentuating the collective. Making spaces have become wildly popular in many types of learning contexts including storefronts, libraries, community centers, museums, and schools (Honey & Kanter, 2013). With access to cheaper tools, the phenomenon of individual hobbyists tinkering in their garages or basements has morphed into groups of people coming together to share tools and approaches—from crafting (e.g., screen printers) to artisan (e.g., jewelry making tools) to advanced digital technologies (e.g., laser cutters). Many educators have fallen in love with the potential that making has to offer.

In part, people are excited by making because by making things, people often use ideas and practices from engineering, mathematics, and science all while attending to aesthetic or artistic aspects of design. STEM fields have long been recognized as some of the higher-paying career paths for the middle and creative classes. The popularized argument is that making is essentially a form of engineering and design that will lead to STEM learning, because students want to make and being successful in making requires that they develop skills and confidence that will support them in deeper future learning activities. Then, that STEM learning will in turn lead to good jobs with high-paying wages. Yet, this uncritical approach to making across learning contexts, including why and how learners engage, will only reproduce existing dynamics around who is served and who is marginalized. STEM educators and others will experience the same exclusions and marginalizations for women and people of color that we have now. By examining making through a literacies frame, we can both expand our understanding of youth and adult practices with representations in ways that highlight and celebrate forms of STEM learning in making that may otherwise be overlooked or dismissed.

Making Spaces

Making spaces are social spaces where individuals and small groups of people design and make objects for personal or commercial use. The reader may have heard of makerspaces and indeed that word is in the title of our book to encourage familiarity with the content. However, in the book itself we use the term *making spaces*. While we appreciate the cultural capital of the term *makerspaces*, we also understand its usage as associated with certain kinds of making activities (e.g., digital fabrication) engaged in by a select group of people, namely middle-class white males such as ourselves (Buechley, 2013; Chachra, 2015). From here on, we use the term *making spaces* to broaden what counts as making activity and to decenter white men as the prototypical makers. In part, using this term is the first step toward beginning to address the critique that the field needs to make explicit how it addresses issues of equity and access in STEM and in making.

In our study, participants raised issues with the term *makerspace*. One said, "What is a makerspace? My makerspace is my kitchen." And another spoke about the ways in which popular conceptions of "making" can reinforce gender roles and devalue the contributions of women. Some of our more experienced interviewees identified with mainstream making and makerspace terminology. But about half of our participants did not. Thus, *making spaces* is part of a crucial conversation in the literature about equity and is reflective of the conversations we had with makers in this study.

Without critical attention to who is supported in making spaces, the kinds of practices and values that educators and learners instill around making, and the attention to the kinds of representational practices that makers use to create objects, we are in danger of reproducing the histories of exclusion for women and people of color that have been perpetrated in STEM learning and American schooling since their inception (Norris, 2014; Calabrese Barton & Tan, 2018).

Our work connects to equity work particularly in making and engineering, but also in science, mathematics, and literacy. In making, Vossoughi and colleagues have questioned the ways in which making itself, and learning in making, are positioned in the field, especially for youth from underrepresented groups (Vossoughi & Bevan, 2014) and the ways in which moment-to-moment learning interactions in a tinkering studio can promote agency among youth of color (Vossoughi, Hooper, & Escudé, 2016). Joining these scholars, we argue for a broader conceptualization of making.

Likewise, Calabrese Barton and Tan have produced a body of work with their colleagues that examines STEM-rich making in after-school activities with youth from historically marginalized communities. They argue that youth community ethnography is a way in which to promote agency and involvement (Tan & Faircloth, 2016) and a way to reposition youth as knowledge producers whose intellectual resources travel across the spaces of their lives (Calabrese Barton, Tan, & Greenberg, 2017). They call this "mobilities of criticality." Youth community ethnography echoes the work we observed at the out-of-school site that serves young people. In that site, we observed youth engaged in technology design projects that met a need they identified in their communities. The STEM-rich making that we observed these youth doing was situated within their lives, happening in the making space but addressing issues and problems they identified in their neighborhoods, schools, and the places they lived.

Meshworking, introduced briefly in Chapters 3 and 4, expands the idea of mobilities of criticality to account for structures of information flows, or the possible lines of communication in people's lives as they move from space to space. Meshworking is an acknowledgment of the distributed expertise of learners' lives and their resources for identifying, organizing, and integrating information across sources (one of the literacy practices we name in this book). An emphasis on meshworks also builds on the idea of "relational equity," explained by DiGiacomo and Gutierrez (2016) as when those people previously thought of strictly in the roles of teachers or learners assume more blurred and egalitarian roles, teaching each other and respecting each other's knowledge base. Taken together, mobilities of criticality, relational equity, and meshworking recognize powered differences as learners travel through different spaces and communicate with others. We draw on studies employing funds of knowledge as a framework for understanding the considerable resources youth of color bring to bear in educational settings (Civil, 2016; González, Andrade, Civil, & Moll, 2001), and how those funds operate within the meshworks of making spaces. Finally, as we describe the different making spaces that serve as the contexts for the literacy practices, we explain how features of those contexts contributed to or impeded literacy. Meshworking as a pedagogical approach attempts to help students build networks of reciprocity and relational equity across those powered spaces, where their funds of knowledge are integral and valued.

There are a few different angles on underrepresentation, equity, access, and transformation that we will be taking in the book. First, we will describe the settings of each of the spaces and the ways in which they do and do not address those issues. For instance, one of the adult making spaces was largely a space populated by white people. While they were very proud of the gender ratio, most folks were not concerned about the racial/ethnic population disparity.

Another out of school space practiced what they called "technologies of the heart/technologies of the earth" in which they unified technological innovation with social/human innovation. This latter move helped to create a caring community that supported youth of color in learning by making through building relationships to selves, peers, and communities. Our implications chapter will also include information on how literacies can be supported in such spaces in ways that contribute to equity for students from underrepresented groups including ways teachers and educators can configure classrooms and learning spaces.

We aim to illuminate the ways the STEM practices that are valued in schools are evidenced in making. We also seek to re-describe the ways these practices are understood in other contexts. We are committed to descriptions of practice that broaden the scope of who is considered a valued participant in these activities. We understand many of the adult and youth practices in making as STEM practices, achieving the same kinds of goals that those practices outlined in formal curricular frameworks are trying to achieve. Our framework is one that builds toward equity by positioning the diverse literacy practices in making as legitimate in STEM.

Chapter 5 introduces the idea and design principle of representational fluidity. Making encourages representational fluidity, or the way learners create and move across representations. In doing so, it acknowledges the validity of different representations in different settings, and it blurs boundaries between informal and formal representations. For instance, makers do not usually need fully drawn schematics to begin their work. Allowing learners to create their own representations gives them authority over their work and the ways in which it is presented. Such positioning is important for learners from communities who have been historically marginalized, who may invent and use representations not typically valued in STEM.

Both schools and making spaces need to address issues of gender, racial, and economic equity. Many see making in schools as a way for students to explore, to find their own passions and be prepared academically through connections to STEM and other subjects (e.g., Bevan, Petrich, & Wilkinson, 2014). With such preparation, they then might become the innovators of the future. We agree. However, learning in making spaces and putting making spaces in more formal institutions, needs to recognize the values and practices of both spaces that serve to reproduce inequity. Yet, making spaces and school spaces do not share many of the same values around STEM practices. They are largely at odds, and we attempt to address that in this book.

There are several differences in relationships to knowledge that do not allow one to simply plop tools or a space into traditional school settings and expect the future of learning to arrive. Making spaces are different from schools and other forms of top-down learning because knowledge is shared, not told. In making spaces, every maker can offer something of value, every maker comes with knowledge that is important to the work. Such a shift in relationships to knowing and learning can be hard for educators who work in institutional settings with very different value structures. Highlighting literacy practices in the context of design and making can be a bridge for teachers and informal educators alike who are looking for a way to make sense of new practices and new approaches to knowledge.

Literacy practices can also be an important entryway to documenting, describing, and understanding learning in these spaces. Participants may create vastly different projects during their time in making spaces, and so it may be hard to document what it is people have learned from one project to the next. A great deal of what makers do is work with tools that certainly involve literacies specific to those platforms. However, representational practices (e.g., talking, reading, writing, and drawing) for various purposes cut across different projects. We study making spaces, and not simply making activities, because it is the *communities* in which making happens that mediate *how* it happens. No making space is quite like any other. There are too many variables: size, tools, people, geographic location. However, we have identified some shared values that we think contribute to literacy learning.

Characteristic Values of Making Spaces

Making spaces are designed to foster creativity, exploration, experimentation, innovation, and learning in STEAM disciplines, particularly design and engineering (IMLS, 2012). One can make anything at Artisan's Asylum, one of the spaces in our book, but it is more than just a room of tools, it is a community of people who have developed shared values, social practices, and knowledge. With that in mind, we use an analytic frame that posits learning is a social and cultural activity that can be analyzed by documenting and assessing what people do in naturalistic environments (Bell, Tzou, Bricker, & Baines, 2012; Cole & Engeström, 1993; Nasir, Rosebery, Warren, & Lee, 2006). It helps us to focus on making spaces as incubators of creativity with certain values and characteristics rather than merely places where individuals engage in interesting work. We start then, with a look at what those characteristic values are in making spaces. Understanding the characteristic values of making spaces has helped us to understand how and why people engage literacies while working, playing, or exploring in them.

We present this as a list (Table 0.1), but it is really a constellation of connected characteristic values in making spaces that are always in play, foregrounded at some times more than others. We suggest these characteristics structure possibilities for developing repertoires of STEM literacy practices. We found these values to describe making spaces across the literature and they emerged from the spaces where we worked. Our list is no doubt lacking the characteristics of some spaces and overly general in others. They are the values we felt were most salient to literacy practices in our studies.

Again, we acknowledge that these characteristics do not exhaustively describe making spaces, nor do they describe each and every unique making space. We use them as an analytic lens to understand the values of making spaces and thus the kinds of STEM literacies that may be important both in negotiating such spaces successfully and in bridging making space activities with disciplinary STEM concepts and practices. For instance, in Chapter 3, we detail how the physical and social spaces of Artisan's Asylum promote values of reciprocity, transdisciplinarity, hybridity, orienting the work toward learning, and distributed expertise. These values have led members to keep walls around their personal studio spaces low, and they contribute to an inevitable culture of interrupting in which members will stop someone else in the middle of a work session to get feedback or help on their own project. Face-to-face communication practices are tied up in the values of the space. It is the communication practices within the context of those values and activities that we consider to be literacy practices, not merely talking to one another, speaking and listening. We will detail this approach to literacies in Chapter 1.

In writing this book, we have tried to map the study and practice of literacies in making spaces. It is new territory, but has yielded rich theoretical and methodological trajectories for understanding and supporting making, literacy, and STEM. These trajectories are emergent. As studies of making, and learning in making, begin to mature, it is our hope that our research will help other educators take up these ideas in their inquiries and designs.

Making Spaces in This Book

The case studies that appear in this book come from spending time in four different making spaces over the course of three years. The adult spaces were Artisan's

TABLE 0.1 Characteristic values of making spaces

Creativity	*Participants' work is driven by an interest in (1) making something new, (2) personalizing common objects, or (3) refashioning something old (hacking), for functional or aesthetic purposes of personal and social value that are different than for what the object was originally intended.*
Cooperation	*People in making spaces work together toward common goals. A common studio space allows people to work together with fewer boundaries than in individual ones. Making spaces serve the purpose of democratizing making in that people or groups of people working together can create works that individuals would not be able to do alone (Artisan's Asylum, 2013; Blikstein, 2013).*
Open sourcing	*Free sharing of ideas, tools, and resources. Relatedly, community and local resources are also valued, similar to "local food" movements (Hatch, 2014). Open sourcing further helps to democratize production by putting it in the hands of individuals through no/low-cost shared access to expensive or high-tech tools for everyday people (e.g.,Buechley, 2006; Gershenfeld, 2007).*
Multimodality	*Throughout the book we give examples of the kinds of materials and representations used in making spaces. Representations include the following: oral and written languages, including highly technical varieties; gestures; 2D and 3D drawings; 3D shapes. Although professional or experienced makers may enter the space knowing how to work with one type of tool or in one modality, they often join spaces because they want to learn new techniques, ways of working, and new representational practices.*
Hybridity	*Making spaces "share some aspects of the shop class, home economics class, the art studio, and science labs. In effect, a making space is a physical mash-up of different places that allows makers and projects to integrate different kinds of skills" (Dougherty, 2013, p. 9).*
Transdisciplinarity	*Making spaces encourage interactions between disciplines and moving beyond disciplinary practices by focusing on the problem to be solved, and the work created (Honey & Kanter, 2013). Whatever is needed is used, and makers are open to thinking outside the box for solutions.*
Networked/Distributed Expertise	*Expertise is distributed across communities of interest. Everyone is expected to know something and no one is expected to know everything (Jenkins, Purushotma, Weigel, Clinton, & Robison, 2009). The interactive capability of the web allows those teaching and those learning to talk to each other across vast geographic and social distances. The wide-reaching character of the web often means that makers draw on multiple sources of information, integrating them into their own projects.*
Community Oriented/ Reciprocal	*Social and distributed problem solving. Members of making spaces often have responsibility for organizing and keeping up the space. Members are expected to be able to offer help as well as receive it. Making spaces are also often connected to the outside communities in which they operate through projects or events.*
Learning Oriented	*Making spaces are "sandboxes," where complex networks of people, tools, materials, and processes intersect and interact in ways that constantly support and promote learning (Halverson & Sheridan, 2014). They provide "a wealth of continuing education in non-standard processes," as a fundamental feature of the space (Ecco, Artist at Artisan's Asylum).*

Asylum and LMNPO Technologies, both in Somerville, Massachusetts. The youth informal space was the South End Technology Center in Boston, Massachusetts. The final space was Nedlam's Workshop in Gateway City (pseudonym). These spaces were chosen because they represented a range of different kinds of making spaces with different kinds of people in them and also because they were ones with which we had connections. Although we describe each space more expansively in the chapters that follow, we offer a brief sketch of these spaces below.

Artisan's Asylum

Artisan's Asylum is an approximately 40,000 square foot, 24-hour, member-run making space that includes many different workshops. Established in another location about four and a half years before we began our observations, it had about 400 active members during our time there. Some of the members come only to use the tools and then leave again. Others rent anywhere from 25 to 100 square feet of studio space where they do much of their work. There are several workshops including: electronics and robotics, jewelry, bicycle shop, metal casting, woodworking, welding and plasma cutting, computer controlled routers, hand and power machining tools, digital fabrication, and a computer lab with professional engineering software installed on each of 15 computers. People run businesses, provide pay-services for other members, and a bicycle gang calls the Asylum their home base. Members pursue their hobbies and teach classes. A few people make their livings primarily by working for others at Artisan's Asylum. There are only three paid staff, so members pitch in to keep the community running, to make infrastructure decisions, to stock the store with Oreos or the electronics lab with LEDs, to fix the machines, and to take out the trash. The population is largely white although there is significant diversity in gender, sexual orientation, and folks who do not or cannot fit in an office job—who might "need to hibernate for a few months in the winter time," as the director of member services told us. We mostly followed long-time (over a year), day-time, full-time members. There were plenty of others that inhabited the place at night, on weekends, or only on particular nights of the week or month with whom we did not interact.

LMNPO Technologies

LMNPO Technologies was an Artisan's Asylum spin-off start-up. Their mission was to provide physical prototyping and small-scale manufacturing services to entrepreneurs and startups. The CEO told us that they wanted to fill the market for people who wanted to produce 5 to 5,000 units of an item. Although there were three original founders (M, N, and P), who all met at Artisan's Asylum, only two were active in the company while we observed them over five months, and one of the principles was seen by our project team only a handful of times. Two of the founders had met as welding teachers at the Asylum, both classified as journeymen in the trades. In addition to the two principles, a creative director, shop manager, and lead designer, the firm employed three shop floor workers and a part-timer who may or may not have been getting paid for his work. LMNPO had many of the same tools as Artisan's Asylum: a three-axis computer-controlled router, a vacuum former, and welding tools. Some of the projects were under non-disclosure agreements and so we could not sit in on meetings about those efforts, but we did observe them produce the mechanical and digital components of an interactive children's museum exhibition, a large touch tablet the size of a table aimed at children with autism, a high pressure plastic injector for molds, and a highly technical coffee pourer for a local high-end coffee roaster. They also had a contract to assemble printed circuit boards with a large electronics manufacturer. The number of contracts dried up as we were observing them and they shut down operation after we had observed them for five months and they had been in business for about two years.

South End Technology Center

The South End Technology Center (SETC) is a neighborhood community center in Boston, MA. It was started by Mel King, a longtime community activist, former state representative and retired MIT professor. While the center provides free training on using computers and video and music production, we focused our research on the work of two related youth programs. The Fab Lab Stewards, which ran from fall to spring, and the Learn 2 Teach, Teach 2 Learn program, which ran from spring to summer. The Fab Lab (Fabrication Laboratory or Fabulous Laboratory) was the first one in the world and sponsored by the Massachusetts Institute of Technology's Center for Bits and Atoms (see www.fabfoundation.org/). It houses a number of computers, four 3D printers, two laser cutters, a vinyl cutter, a 4' x 8' computer-controlled router, and various other small woodworking and electronics tools. Two days a week, anyone is allowed to walk in and make something with the supplies and tools at the Fab Lab. The Fab Lab Stewards are high school youth employed to run the lab, teach people who come in, and assist them with projects. In the summer, the community lab hours stop and the space is taken over by the Learn 2 Teach, Teach 2 Learn program that employs about 40 youth to a) learn about different digital fabrication tools and b) teach introductory classes to younger children at the space and at camps around the city.

Nedlam's Workshop

This 3,500 square foot former woodshop is housed in Gateway City's High School, a large urban comprehensive school just outside Boston. It was started with prior National Science Foundation support (IIS-1450985, PI Brian E. Gravel) for the purpose of exploring the role of "engineering inquiry" (i.e., making) in how school communities construct values around learning and knowledge production. The space houses a large range of woodworking equipment (e.g., table saws, band saws, planers, sanders, scroll saws, lathes), a laser cutter, two 3D printers, a CNC mill, an electronics bench, a "fabric arts" studio, LEGO robotics equipment, and various raw materials, hand tools, and left over bits and pieces. The center of the shop is a collection of large 4'×4' tables for shared project work. Nedlam's Workshop was opened in October 2014, and at first hosted after school "open studios" where youth were invited to make things that interested them. In the spring of 2015, the first iteration of *Creative Design and Engineering* was designed and taught by a team of researchers from Tufts University. Following that experiment, Nedlam's Workshop has hosted professional development workshops, summer programs for youth, after-school open studios and workshops (e.g., a series of e-textile workshops). All of this work focused on engineering inquiry and supporting teachers and youth to identify "pressure points" in their lives—places where the world is oppressive, inhospitable, or could simply be more enjoyable—and using the tools of making and engineering to design and make solutions to those problems.

Our work there with teachers at Gateway City High School began in the Fall of 2016. We offered professional development workshops to mathematics and science teachers focused on STEM literacies in making. We also worked closely with a physics and engineering teacher to redesign the *Creative Design and Engineering* course. We participated as facilitators and observers in that course, working closely with the teacher and the students. Simultaneously, we held bi-weekly seminars with teachers from the PD to continue supporting their thinking about STEM literacies in their classrooms. We helped them design different activities to play with these ideas in their classrooms, and we observed implementations in the late spring.

The Structure of the Book

We have organized the book as a progression of case studies focused on the sites we studied, the literacy practices we observed, and theoretical proposals that advance the field's understandings of how representations are produced and used in making spaces. We work from descriptions of literacies in adult-oriented making spaces, move on to complicate those descriptions by examining youth practices, and finally explore specific literacies across spaces to compare and contrast the ways representations functioned in different tasks. Chapter 1 introduces the STEM Literacies in Making Spaces framework and describes the settings in this study. From there, each chapter focuses on one particular literacy practice, elaborating it with evidence from our observations and designed interventions. Chapter 7 focuses on how teachers learn in making spaces, and that leads into the final chapter where we discuss the implications of this research. In that final chapter, we introduce design principles developed from the STEM Literacies in Making Spaces framework, relate those to specific implications for teachers, and finally make recommendations for areas of future research at the intersections of literacies, STEM, making, and equity.

Next, Chapter 1, lays out a cohesive theoretical framework for the case studies included in the volume. It describes why we conceptualize literacies as *multiliterate social practices*. It describes what we mean by STEM literacies in particular and names what we have identified as STEM literacies in making spaces.

References

Artisan's Asylum. (2013). Retrieved October 12, 2013 from http://artisansasylum.com/?page_id=2555

Bell, P., Tzou, C., Bricker, L. A., & Baines, A. D. (2012). Learning in diversities of structures of social practice: Accounting for how, why and where people learn science. *Human Development, 55*, 269–284.

Bevan, B., Petrich, M., & Wilkinson, K. (2014). Tinkering is serious play. *Educational Leadership, 72*(4), 28–33.

Blikstein, P. (2013). Digital fabrication and "making" in education: The democratization of invention. *FabLabs: Of Machines, Makers and Inventors, 4*, 1–21.

Brown, A. L. (1992). Design experiments: Theoretical and methodological challenges in creating complex interventions in classroom settings. *The Journal of the Learning Sciences, 2*(2), 141–178.

Buechley, L. (2006, October). *A construction kit for electronic textiles.* In 10th IEEE International Symposium on Wearable Computers, 2006 (pp. 83–90).

Buechley, L. (2013). *Thinking about making.* In Keynote presented at the Annual FabLearn Conference, Palo Alto, CA.

Calabrese Barton, A., & Tan, E. (2018). A longitudinal study of equity-oriented STEM-rich making among youth from historically marginalized communities. *American Educational Research Journal, 55*(4), 761–800.

Calabrese Barton, A., Tan, E., & Greenberg, D. (2017). The makerspace movement: Sites of possibilities for equitable opportunities to engage underrepresented youth in STEM. *Teachers College Record, 119*(6), 1–44.

Chachra, D. (2015, January 23). Why I am not a maker. *The Atlantic.* Retrieved from www.theatlantic.com/technology/archive/2015/01/why-i-am-not-a-maker/384767/

Civil, M. (2016). STEM learning research through a funds of knowledge lens. *Cultural Studies of Science Education, 11*(1), 41–59.

Cole, M., & Engeström, Y. (1993). A cultural-historical approach to distributed cognition. *Distributed Cognitions: Psychological and Educational Considerations*, 1–46.

Collins, A., Joseph, D., & Bielaczyc, K. (2004). Design research: Theoretical and methodological issues. *The Journal of the Learning Sciences, 13*(1), 15–42.

DiGiacomo, D., & Gutiérrez, K. (2016). Relational equity as a design tool within making and tinkering activities. *Mind, Culture & Activity, 23*(2), 1–15.

Dougherty, D. (2013). The maker mindset. In M. Honey & D. Kanter (Eds.), *Design, make, play: Growing the next generation of STEM innovators* (pp. 7–11). New York, NY: Routledge.

Tan, E., & Faircloth, B. (2016). "I come because I make toy": Examining nodes of criticality in an afterschool Science and Engineering (SE) club with refugee youth. In *Qualitative research in STEM* (pp. 44–67). Abingdon: Routledge.

Gershenfeld, N. (2007). *Fab: The coming revolution on your desktop—from personal computers to personal fabrication*. New York, NY: Basic Books (AZ).

González, N., Andrade, R., Civil, M., & Moll, L. (2001). Bridging funds of distributed knowledge: Creating zones of practices in mathematics. *Journal of Education for Students Placed at Risk*, *6*(1–2), 115–132.

Halverson, E. R., & Sheridan, K. (2014). The maker movement in education. *Harvard Educational Review*, *84*(4), 495–504.

Hatch, M. (2014). *The maker movement manifesto: Rules for innovation in the new world of crafters, hackers, and tinkerers*. New York, NY: McGraw Hill.

Honey, M., & Kanter, D. (2013). *Design, make, play: Growing the next generation of STEM innovators*. New York, NY: Routledge.

Institute of Museum and Library Services (IMLS). (2012, September). *Talking points: Museums, libraries, and makerspaces*. Washington, DC. Retrieved October 12, 2013 from www.imls.gov/assets/1/AssetManager/Makerspaces.pdf

Jenkins, H., Purushotma, R., Weigel, M., Clinton, K., & Robison, A. J. (2009). *Confronting the challenges of participatory culture: Media education for the 21st century*. Cambridge, MA: MIT Press.

Nasir, N., Rosebery, A. S., Warren, B., & Lee, C. D. (2006). Learning as a cultural process. In K. Sawyer (Ed.), *Handbook of the learning sciences* (pp. 489–504). Cambridge: Cambridge University Press.

Norris, A. (2014). Make-her-spaces as hybrid places: Designing and resisting self constructions in urban classrooms. *Equity & Excellence in Education*, *47*(1), 63–77.

Papert, S. (1980). *Mindstorms: Children, computers, and powerful ideas*. New York, NY: Basic Books, Inc.

Vossoughi, S., & Bevan, B. (2014). Making and tinkering: A review of the literature. *National Research Council Committee on Out of School Time STEM*, 1–55.

Vossoughi, S., Hooper, P. K., & Escudé, M. (2016). Making through the lens of culture and power: Toward transformative visions for educational equity. *Harvard Educational Review*, *86*(2), 206–232.

Vygotsky, L. S. (1978). *Mind in society*. Cambridge, MA: Harvard University Press.

1

STEM LITERACY PRACTICES

Making activities are replete with literacies, particularly as they relate to Science, Technology, Engineering, and Mathematics (and Arts) fields (STEM or STEAM) across professional, school, and informal making spaces (Gravel, Tucker-Raymond, Kohberger, & Browne, 2017). This chapter outlines our framework for identifying, understanding, and supporting STEM literacy practices in making spaces. We explain what literacy practices are and why we use them as a focal unit of analysis. Next, we present our STEM literacy framework that we used to begin our analyses of the spaces.

What Are Literacy Practices?

Many people think of literacy as the ability to read and write written language. This book treats literacy more broadly. We define literacy to be one's use of, and facility with, representations, or texts, to make meaning and communicate with others. It includes both consuming (e.g., reading) and creating (e.g., sketching) with a variety of representations. Second, we understand there to be multiple literacies: for instance, knowing how to write a legal brief for a court proceeding is one literacy. Creating an attractive and easily navigable website is another. There are diverse literacies in use for diverse purposes in making spaces. To help us understand STEM literacies in making spaces we build on extensive work in literacies scholarship that foreground *social practices* (Heath, 1983; Street, 2003) and multiple literacies, or *multiliteracies* (e.g., Cope & Kalantzis, 2000) to understand and support learners in multiple contexts. Taken together, we make the argument that literacies are best understood as *multiliterate social practices*.

Social Practices

This book is not about how individuals engage in making. It is about what happens in making spaces. The focus on spaces necessitates our analytic lens at the level of social practice. A social practice perspective describes and analyzes how individuals and communities relate to one another and organize, reproduce, and transform ways of living together (Cole, 1996; Rogoff, 1995). Within a social practice framework on literacy (Barton & Hamilton, 2000) people work together in social groups to produce representations that communicate ideas and act as tools to think with (Vygotsky, 1978). What people learn is a part of the valued institutional and community practices involved in the contexts in which they are participating. People also draw on their own "repertoires' of practice," or ways of engaging in activities, drawn from past experiences in multiple communities, each with their own traditions (Gutierrez & Rogoff, 2003). Through literacy practices, people communicate what knowledge is valued in a particular social and cultural setting as they use and create shared repertoires for meaning making and frameworks for interpretation (Dyson & Genishi, 2005).

Literacies are contextually constituted by the purposes, identities, and tools, including ways of communicating, valued in that particular place and time (Gee, 1996; Street, 2003). One can think of literacies as occurring at different grain sizes: events, practices, and ideologies and structures. Literacy events are time- and space-bound interactions in which one or more multimodal texts has a central role (Heath, 1983). For instance, an individual subway commuter surfing the Web on their phone while on the way to work Monday morning is a literacy event. Literacy practices connect literacy events with broader social and cultural values about reading, writing, and communicating (Barton & Hamilton, 1998) in ways that are mediated by local values and purposes. Surfing the web on the way to work becomes a literacy practice because it is something that this commuter, and many others like her, do regularly. The infrastructure that has enabled wifi or cellular service on subway cars, and thus the practice of surfing the web while moving in a crowded container through underground tunnels, is a part of larger ideologies and structures, ones that assert people need to be connected to information at all times.

Thinking of and approaching our understanding of literacy at the level of situated practices allows us to make connections between what people do in making spaces, the ways in which they use representations, and the larger value structures surrounding and embedded in the spaces they work. It enables us to connect the makers' use of representations to the values, ideologies, political and economic structures, and physical spaces of the making space and the larger social milieu of which it is a part. Literacy practices, then, allow us to see the ways in which people communicate and make meaning through texts in different spaces with different ideologies and different relationships between knowledge and power. Methodologically, it helps our analysis to understand powered relationships between spaces and practices, how literacy practices are similar and different in different spaces, and how different parts of participants' other experiences and communities come into play. As part of pedagogical design, understanding literacies as practices helps educators and learners understand how, why, and when texts are produced and consumed and by whom. Because there are social norms and rules about who can produce and read texts, limiting or giving access to information, literacy is intimately tied to power. Supporting learners' literacy practices helps give them access to that power. In the next section, we theorize the multiplicity of literacies as multiple.

Multiliteracies

Different practices, values, purposes, languages, and modalities lead to different kinds of literacies. The contemporary connected world requires that individuals be literate in multiple domains. Literacies manifest through "increasing local diversity and global connectedness" (New London Group, 1996, p. 64). There is not one way to be literate. Every community values different knowledge, different relationships to power, and different forms of communication. Because different communities co-exist and intermingle with one another, these forms also co-exist and require people to be literate in multiple ways (Cope & Kalantzis, 2000). Multiple literacies, or *multiliteracies* accounts for literacy practices as a part of cultural and institutional practices that differ in different communities (Scribner & Cole, 1981; Gutiérrez, 2008); as having differentially powered relations to texts (Lewis, Enciso, & Moje, 2007); as existing in multiple modalities (Kress, 2010; Rowsell, 2013), and we argue, it supports an understanding of literacy practices as linked to intersectional identities (Kirkland, 2013; Winn, 2010). We discuss the central dimensions of multiliteracies, including attention to the historical production of literacy practices, multimodality as it relates to making spaces, shifting landscapes of digital media production, and the tightly knit relationship between multiliteracies and identity.

Earlier, we discussed how literacies are part of social practices. As social practices in communities, literacy practices are differentiated by purpose and stratified by value. Communities are also historical entities. To maintain continuity and cohesiveness, they rely on practices that have been reproduced through time, including what one does with texts. Different literacies are valued in making spaces than are valued in schools. To learn the literacy practices of a given community is to give one access to cultural and symbolic capital that one can use to establish one's legitimacy in a community, with a history and its own social norms (Bourdieu, 2003). In the case of this book, we are concerned with making spaces as institutions in their own right, making spaces as part of schools, and schools as institutions in their own right.

A second part of multiliteracies is that there are multiple kinds of representations. Representations can take any form as long as they are symbols that mean something to somebody. Those representations include written or spoken language; mathematical notations; images both moving and static, photographed and drawn; annotated schematics; design software; and gestures (Kress, 2010; Rowsell, 2013). What is more is that these representations can exist together, such as the words, images, and videos found on a website or a microchip schematic with electrical symbols, notations, and drawings of circuits. The characteristic of multiple representations is known as multimodality and helps to understand the breadth of representations that people create and use in making spaces. Different spaces, such as making spaces and schools value different configurations of modalities. It is important for makers using electronics to be able to read technical specification sheets (spec sheets) that contain written language, abbreviations, symbols, diagrams, and technical attributes of a given product. Learning to read a spec sheet is not a valued practice in K–12 schools. Instead, schools often value written and spoken language that explains or argues.

Perspectives on multimodality in literacies increasingly draw on work in digital domains (Lankshear & Knobel, 2008; Kress, 2003; Mills, 2010). Of particular interest to our work are the studies that focus on multimodal production (Kress, 2010; Miller & McVee, 2013), remixing and digital media production (Erstad, 2008; Peppler & Kafai, 2007), and participation in social media networks (boyd, 2014; Knobel & Lankshear, 2008). Positioning learners as producers of multimodal texts transforms their relationships to representations of knowledge, from knowledge consumers to owners of that knowledge. Remixing and digital media production rely heavily on connected networks of wide access to open source information and resources of social media and other platforms. Digital literacies not only make it possible to produce digitally fabricated objects, but also connect the worlds of do-it-yourselfers together so that people can learn from one another, no matter how far apart they are. It makes it necessary to understand literacy practices in making spaces as related to other places and spaces in the world. What places and spaces, and what sources of information making spaces are connected to, can indicate the social, ideological, and relational commitments of different making spaces. One of the most important contributions of scholarship on digital literacies is the notion that what is required to be literate is always changing because forms, relationships, and purposes are always changing (Leu, 2000).

Lastly, a theory of multiliteracies recognizes that literacy practices are tied to people's identities (Moje, Luke, Davies, & Street, 2009). The literacy practices that people engage, the informal and formal languages that they use, the technicality of their drawings, and their gestures all constitute and signal self-, assigned-, and imagined-identities. Identities are representations of types of people, recognized by others in social settings (Carlone & Johnson, 2007). As socially negotiated representations, identities become a part of symbolic capital.

Multiliteracies also recognizes that people participate in literacies and identity work in multiple places in their lives (at home, at school or work, in church, during sporting events, or political actions) and bring these literacies and identities with them to the making space.

Thus, the literacy practices of making spaces are mediated by what else people know how to do and how else they present themselves, and are recognized, as kinds of people. Making spaces, and their practices, are permeable. Their shape is subject to the people who move in and out of them each day (Massey, 2005). However, their institutional histories and relationships to particular literacies and identities also shape who participates. A making space may foreground work on electronics for some members, but when those same electronics are embedded within a focus on hip hop music production, we expand the entry points for different literacies and identities and invite new members to participate.

In this book we are most concerned with those multiliteracies related to STEM fields. In schools these have been classified as disciplinary literacies (e.g., Moje, 2007, 2015; Shanahan & Shanahan, 2008), or as encompassing the literacy practices of certain subject areas in school and their relations to fields of study (e.g., mathematical literacy). In making spaces, the disciplines are not so clearly defined, and practices exists within and across intersections of different disciplinary traditions. For instance, nearly everything involves some kind of technology or tool, digital or not. In the next section we discuss what we mean by STEM literacies.

What Are STEM Literacies?

There has been a recent focus on disciplinary literacies in studies of schooling (Manderino & Castek, 2016; Moje, 2015). By disciplinary literacy, we mean more than subject matter, more than just knowing how to read a science textbook or follow the steps of the engineering design process. This perspective positions disciplines as sociohistorically situated and constantly negotiated, producing their own kinds of practices, including the ways in which they use texts (Bucciarelli, 1994). Disciplines have historical ways of doing things and literate habits of mind (Moje, 2007; Wineberg, 2001). In science for instance, literate habits of mind might include documenting results of controlled experimental trials, or learning how to locate quality sources of information for different topics.

We agree to some extent that literacies are disciplinary, for example, knowing how to make a computer aided design (CAD) drawing is particular to engineering and design, and CAD tools have embedded symbols and language that have been inscribed by the discipline over the years. We also argue that the practices surrounding particular uses of texts can be applied more generally across disciplines. Sketching is a practice that people can do in many different disciplines as a way to think through an idea or imagine a system of interrelated parts. However, rather than treating each discipline as separate, as in school, we believe STEM is actually a better disciplinary frame than any single of the academic disciplines. Using STEM allows us to honor different disciplines and, at the same time, recognize that there are powerful practices that bring together elements of different ones.

Because making is project based, it is inherently multidisciplinary. The activities we observed all had components of science, technology, engineering, and mathematics. Many had their primary goal as learning some dimension of content and/or practice in any combination of those fields, for example, learning how to enact sensing-control-actuation paradigms (i.e., robotics) as components of interactive public art pieces requires knowledge of computer programming, electronics, kinematics, mathematical modeling, cognitive psychology, public response, and probably much more. It also requires knowing how to read and translate different representations, look for information, talk to others to give and receive feedback, create representations that others can understand, document what one has done for future testing and troubleshooting, and share the work with others once it is finished.

STEM is not in every making project, but one can find it in most. For instance, the jewelry makers as Artisan's Asylum need to know about the properties and characteristics of different metals. They need to know not to use copper when working with gold because it will oxidize the surface. They need to know what chemicals can remove the pink tint, and how hot to keep their flames when melting and forming shapes. They need to know about scale, proportion, and surface area. They need to know how to source materials. They might also want to know how to create molds from 3D printed materials. The point is, making does not fall squarely in any one STEM discipline. At the same time, STEM is used in service of many making projects, even if engaging in engineering, or doing math, is not the explicit goal.

Literacy Practices as Units of Analysis

The five basic STEM literacy practices we have identified are these:

- Identifying, organizing, and integrating information across sources.
- Creating representational forms and traversing representational systems.
- Communicating with others for help and feedback.
- Sharing finished or almost finished objects.
- Documenting making processes and/or milestones.

The framework we present in Table 1.1 introduces the definitions of people's STEM literacy practices we identified in making spaces. It is elaborated in the rest of the book by attending to important themes of how literacies work

TABLE 1.1 Definitions of STEM literacy practices

STEM Literacy Practice	*Definition*
Identifying, organizing, and integrating information across sources.	*Searching for information on the Web, using reference guides, material specification sheets, and product manuals* *"Maybe like it's some mixture of finding resources and trying things, and then the resources span from like, textbook to, you know, random people on the Internet and then talking to people, observing things."*
Creating representational forms and traversing representational systems	*Producing symbolic representation of any aspect of the project as an aid to making (measurements, notes, drawings) or transitioning between different modalities such as talk, sketches, and formal designs.* *"Over time it just became part of my practice when it comes to sketching things and developing ideas. Half the time I use a pencil and half the time I use pieces of wood and scraps and paper and whatever else. It's just a way of working out ideas in the physical world."*
Communicating with others for help and feedback.	*Help includes providing or seeking assistance to complete a task or series of tasks that one could not do alone. More solicited, more hands-on, more involvement than feedback. Feedback includes providing or seeking advice, opinion, or consultation. Includes general guidance. Can be unsolicited.* ***Help:*** *"because my sewing knowledge is much less than the other areas, I really have to rely on the people there to basically refresh my memory and teach me things. So pretty much every time I do there I'm always asking different questions or making sure that I'm doing things correctly."* ***Feedback:*** *"Sometimes I'll talk it over with friends and family, whoever is around, just kind of [say] what I'm thinking and see if they have any input to the problems, the roadblocks I'm running into or the general idea."*
Sharing finished or almost finished objects.	*Demonstrating made artifacts, producing web pages or using social media to promote work, and peer-to-peer teaching. Raising or earning money through crowdfunding, or directly selling objects and services.* *"Blogging about it, I'm not as active on the keyboard forums as I mean to be. We have a regular mailing list that we send to our users, to our potential customers, people who've signed up for it."*
Documenting making processes and/or milestones.	*Photographing, writing about, or recording the steps, processes, and output of a project for oneself or as a reference for others.* *"Documenting and then just uploading it on to like a facebook and adding a description, and that sort of to me like encapsulates that process."*

through people's goals, their powered relationships to one another, local values of the space as an institution or as part of an institution, and larger social norms and discourses about making, STEM, and learning. In other words, the studies in the book articulate the literacy practices in the framework within particular social settings, expanding their basic definitions to describe how they are actually instantiated and how they might be identified in making spaces. In what follows we present our basic definitions of the literacy practices we discuss in this book in an attempt to help educators and researchers start to recognize and support literacy practices in their own making spaces, be they 40,000 square foot warehouses or 30 seat classrooms.

Even though we have named the literacy practices individually, we think about them more expansively—as operating on multiple levels, and often in conjunction with each other, depending on the particular tasks/processes. Furthermore, literacies rarely occur by themselves. They are often part of a string of practices that are employed throughout the life of a project or between people involved in entirely different projects but who share the same space. In part, the creation of these five descriptions of the practices was a way for us to find the literacies at the intersections of making, school STEM subjects, and disciplinary STEM fields in ways that supported expansive relationships to knowledge and recognition of different identities,

Building Bridges Between Formal and Informal Spaces

One of the goals of our work is to use literacies in the service of STEM learning as a way to bridge out of school and in school learning. Deep, engaging learning happens out of school. Yet, school remains a place of alienation for many learners, including those from groups underrepresented in STEM fields (African American, Latinx, English language learners). That is, we want to be able to draw on literacies from informal learning settings that create safe, active spaces (Klimczak, Wallace, & Gaskins, 2015) and draw on learners' funds of knowledge (Wilson-Lopez, Mejia, Habun, & Kosun, 2016) and community practices (Calabrese Barton & Tan, 2018).

To begin building the bridge from the school side, our articulation of STEM literacies in making drew from "practices" of the Next Generation Science Standards (NGSS) and Common Core State Standards in English Language Arts (CCSS) documents, where an emphasis on literacies is located even if those practices are not framed as literacies, per se. We started with those sources because we wanted to understand the ways in which makers in informal and professional spaces used STEM in ways that allowed them to have strong connections to the same kinds of practices students are expected to engage in during authentic STEM coursework in schools. Thus, for instance, in the NGSS, it states that, "scientists and engineers present data in a myriad of visual formats in order to reveal meaningful patterns and trends" (NGSS Lead States, 2013,

Appendix M, p. 2). In our construction of STEM literacy practices, we used this definition to help formulate our meaning for sharing finished objects but also for multimodality and the literacy practice of creating and traversing representational systems.

We agree with Quinn and Bell (2013) that informal learning and formal learning spaces are "part of the same ecosystem for [STEM] education in that they share the goal of cultivating the same intertwined strands of [STEM] learning outcomes" (p. 19). Literacy practices, as outlined by a joint committee of NGSS and CCSS development teams, help to make and support connections to conceptual understandings by including,

> an appreciation of the norms and conventions of the discipline of science, including understanding the nature of evidence used . . . the capacity to make and assess intricate arguments, synthesize complex information, and follow detailed procedures and accounts of events and concepts . . . gain knowledge from elaborate diagrams and data that convey information and illustrate scientific concepts. Likewise, writing and presenting information orally are key means for students to assert and defend claims in science, demonstrate what they know about a concept, and convey what they have experienced, imagined, thought, and learned.
>
> *(NGSS Lead States, 2013, Appendix M, p. 1)*

What NGSS and CCSS fail to capture are the structural, institutional, ideological, and local values and practices that make up the roles, divisions of labor, community rules, tools and goals that are a part of every activity (Engeström, 2014). That is, learning goals such as, "make and assess intricate arguments, synthesize detailed procedures and accounts of events and concepts" only make sense within a larger goal-directed activity. That activity in school might take the form of a teacher telling students what topics they will study, what evidence will count, and what side of the argument they might be on. In schools, it is assumed that those doing the arguing are on equal footing. That the merits of the argument, the claims and evidence, will win the debate.

In a making space, that activity might be more organic. As we report on in Chapter 4, a group of people were working on a programmable remote controlled car. A high school male was using a motor controller and was told by his female college mentor, an electrical engineering major, to make sure he was connecting it correctly, and that he might blow it up if he did it wrong—hadn't he heard about the exploding hoverboards? He declined to believe her and, when he connected the battery, shocked himself in the process. The battery soon began emitting smoke. Who wins this argument in this case is not simply based on claims, facts, or evidence as NGSS would have us believe. It depends on relationships between the people arguing and relationships to the others in the room. When the same young man was told by an older male mentor that he could seriously hurt everyone at the table, he took it seriously and decided to find out more about the battery. We use this story to point out that "making and assessing intricate arguments" is a circumscribed description of what really happens in schools, in making spaces, and in STEM fields. The high school student did not listen to his college mentor, despite her experience and citation of evidence. His lack of trust probably had something to do with her age, her gender, and the combative style of communicating that they seemed to share with each other in their history of interactions.

Literacy practices are dependent on who is participating, what their goals are, and their relationships to each other, the community, and the institutional norms and rules. These different configurations create different relationships to knowledge and literacies. Making spaces are different from schools and other forms of top-down learning because knowledge is shared, not told. In making spaces, every maker can offer something of value, but there are still social relationships and identities which one has to negotiate. Such a shift in relationships to knowing and learning can be hard for educators who are used to more teacher-centered classrooms. Highlighting literacy practices, as socially situated, in the context of design and making can be a bridge for teachers and informal educators alike who are looking for a way to make sense of new practices and new approaches to knowledge that making encourages.

STEM Literacy Practices in Making Spaces: Expanding Definitions

The book is organized around literacy practices as the core unit of our investigation. However, the chapters are not solely stories about the literacy practices we present in Table 1.1. Rather, they are case studies and cross-case analyses of the spaces and activities we observed. They showcase how STEM literacy practices unfolded in the making spaces as part of multiliterate social practices. We offer abstracts of each of the chapters, organized to show how they are arranged in such a manner as to build the argument of the importance and utility of a focus on literacies in making spaces. As this is new territory, we offer descriptions as well as some concluding thoughts as the "so what?" of each chapter, while realizing that each of the ideas and literacies outlined in the book can continue to be explored at much greater depths, with much greater complexity.

Chapters 2, 3, and 4 highlight specific making spaces, beginning with an investigation of learning and ***communication*** in a professional study at LMNPO Technologies, a professional engineering firm specializing in small batch rapid prototyping. **Chapter 2** describes how the approach to learning at LMNPO was mediated by numerous factors, including the histories of the firm's owners own learning opportunities and participation in do-it-yourself settings, and the organizational structure of the firm. In particular, we describe the tensions around how shop floor workers were expected to learn new technologies within their work, the forms of support offered through apprenticeship by observation and just-in-time teaching, and the struggles between learning and meeting the needs of the client. These tensions are discussed as critical considerations in youth making spaces, of particular concern to thinking about teaching in making spaces. **Chapter 3** focuses on an adult-only making space called Artisan's Asylum, and we discuss how the physical, cultural, and social space of Artisan's Asylum, a 40,000 square foot making space with multiple shops and about 400 active members, mediates the ways in which people communicate with and learn from one another as they work on individual and collaborative projects. Specifically, we describe the interrupting culture of the Artisan's Asylum, or the ways in which people "interrupt" each other in while they are working to ask questions or seek guidance, and the emergent social and professional networks that result. Chapters 2 and 3 highlight the role of ***communication*** in the formation of these dynamic networks within each community, and how those networks support the flow of information in support of making. **Chapter 4** goes deeper into ***identifying, organizing, and integrating information*** as a literacy practice by focusing on a paid summer work program for youth at a community-based making space, the South End Technology Center (SETC). We explore how youth learn to search for information and materials they need for their projects. In particular, we followed the activity around a communal table at which youth were sitting and the ways in which they used web browsers to search for materials and interact with each other for feedback and socializing during the search. Youth exhibited many of the practices that our expert makers did, yet often positioned themselves as not knowledgeable or in need of guidance as a means of maintaining and leveraging social contact. The chapter expands the notions of networks to consider *meshworks*, a construct that helps to describe social movement, relationship building, and information gathering in making spaces. Taken together, these three chapters examine the literacies in each of these spaces to sketch a view of making as intricately interwoven with the spaces themselves through the modes of communication, social networks, and ways information flows in what we call meshworks. We attempt to illustrate how one makers success is actually the collective product of the individuals of the space, whose interactions, identities, and practices constitute the space itself (Massey, 2005).

Chapters 5 and 6 are cross-case analyses of two specific literacy practices, ***using and creating representations*** and ***sharing both in person and online***. We build from Chapter 4's discussions of the meshwork to explore how knowledge is expressed and shared through the production of representations in **Chapter 5**. Sketches, diagrams, notes, and more formal representations like those from computer aided design that require precise measurements are all central to the work of designers and makers. In this chapter we discuss the ways in which makers create and traverse, or move across, representations in multiple making space contexts. We introduce the concept "representational fluidity" to describe how professional makers and hobbyists embrace the informal nature of their work in the space of Artisan's Asylum to purposely use sketches rather than more formal, technical drawings; how youth at SETC move from sketches, to schematics, to diagramming the electrical circuits they make as an aid in the fabrication of an interactive water piano; and at Nedlam's, the school making space, how students sketch across different materials as they continually frame and understand the problems they are solving. **Chapter 6** describes the ways in which these representations are used by makers to **share their work** on the Web. The chapter presents three different kinds of sharing: a) a youth creating a "how-to" on the website Instructables.com, b) two adult entrepreneurs using Kickstarter to crowdsource financing for their product, and c) youth in a summer program used blogging to share their work-as-inspiration for future program youth. We use these instances to theorize about the role of identities and presentation of self in making literacies.

Finally, we focus on our work with teachers in the third year of the study. **Chapter 7** describes aspects of teacher learning and the ways ***creating and traversing representations*** can help teachers rethink what and how they teach. We use a case study of a workshop our team held with STEM teachers at Nedlam's Workshop to explore how teachers learn in making spaces. It focuses on two mathematics teachers who were making 3D shapes first out of paper and then with a 3D printer and extruded plastic. The chapter discusses how teachers creating new kinds of representations develop insights about mathematics and the tools they used with students. It also discusses the ways in which relationships to students, and values of making spaces, need to be considered when integrating making into STEM classrooms.

To finish out the book, **Chapter 8** draws from each of the case studies to discuss the opportunities and tensions for informal educators and teachers interested in promoting STEM literacies in informal, school-based, and hybrid spaces. It makes explicit connections to STEM subjects (both content and practices) by drawing lessons from the previous case studies and cross-case analyses. We present a set of design principles for the careful treatment of literacies in making spaces, intended to support educators and future researchers. Finally, we discuss implications for future research in making and making spaces as they relate to the consumption and production of representations in social practice.

The **Appendices** describe our methodology and provide some of the literacy scaffolds we designed to support makers. The appendix on methodology describes our sampling and data collection procedures, what was common and different about data collection across each of the spaces, and the ways in which we analyzed data and created case studies as a way to present our findings.

References

Barton, D., & Hamilton, M. (1998). *Local literacies: A study of reading and writing in one community*. London: Routledge.

Barton, D., & Hamilton, M. (2000). Literacy practices. In D. Barton, M. Hamilton, & R. Ivanič (Eds.), *Situated literacies: Reading and writing in context* (pp. 7–15). London: Psychology Press.

Bordieu, P. (2003). *Language and symbolic power*. Cambridge, MA: Harvard University Press.

boyd, d. (2014). *It's complicated: The social lives of networked teens*. New Haven, CT: Yale University Press.

Bucciarelli, L. L. (1994). *Designing engineers*. Cambridge, MA: MIT Press.

Barton, A. C., & Tan, E. (2018). *STEM-rich maker learning: Designing for equity with youth of color*. New York, NY: Teachers College Press.

Carlone, H. B., & Johnson, A. (2007). Understanding the science experiences of women of color: Science identity as an analytic lens. *Journal of Research in Science Teaching, 44*(8), 1187–1218.

Cole, M. (1996). *Cultural psychology: A once and future discipline*. Cambridge, MA: Harvard University Press.

Cope, B., & Kalantzis, M. (Eds.). (2000). *Multiliteracies: Literacy learning and the design of social futures*. London: Psychology Press.

Dyson, A. H., & Genishi, C. (2005). *On the case* (Vol. 76). New York, NY: Teachers College Press.

Engeström, Y. (2014). *Learning by expanding: An activity-theoretical approach to developmental research*. Cambridge: Cambridge University Press. doi:10.1017/CBO9781139814744

Erstad, O. (2008). Trajectories of remixing: Digital literacies, media production, and schooling. *Digital Literacies: Concepts, Policies and Practices, 30*, 177–202.

Gee, J. P. (1996). *Social linguistics and literacies: Ideology in Discourses* (2nd ed.). London: Taylor & Francis.

Gravel, B. E., Tucker-Raymond, E., Kohberger, K., & Browne, K. (2018). Navigating worlds of information: STEM literacy practices of experienced makers. *International Journal of Technology and Design Education, 28*(4), 921–938.

Gutiérrez, K. D. (2008). Developing a sociocritical literacy in the third space. *Reading Research Quarterly, 43*(2), 148–164.

Gutiérrez, K. D., & Rogoff, B. (2003). Cultural ways of learning: Individual traits or repertoires of practice. *Educational Researcher, 32*, 19–25.

Hatch, M. (2013). *The maker movement manifesto: Rules for innovation in the new world of crafters, hackers, and tinkerers*. New York, NY: McGraw Hill Professional.

Heath, S. B. (1983). *Ways with words: Language, life and work in communities and classrooms*. Cambridge: Cambridge University Press.

Kirkland, D. E. (2013). *A search past silence: The literacy of young Black men*. New York, NY: Teachers College Press.

Klimczak, S., Wallace, A., & Gaskins, N. (2016). Technologies of the heart: Beyond # BlackLivesMatter and toward #MakingLiberation. In Blikstein, P., Martinez, S. L., & Pang, H. A. (Eds.). *Meaningful Making: Projects and Inspirations for Fab Labs+ Makerspaces* (pp. 61–67). Torrance, CA: Constructing Modern Knowledge Press.

Knobel, M., & Lankshear, C. (2008). Remix: The art and craft of endless hybridization. *Journal of Adolescent & Adult Literacy, 52*(1), 22–33.

Kress, G. (2003). *Literacy in the new media age*. New York, NY: Routledge.

Kress, G. (2010). *Multimodality: A social semiotic approach to contemporary communication*. New York, NY: Routledge.

Lankshear, C., & Knobel, M. (2008). *Digital literacies: Concepts, policies, and practices*. New York, NY: Peter Lang.

Leu, D. J. (2000). Literacy and technology: Deictic consequences for literacy education in an information age. In M. L. Kamil, P. Mosenthal, P. D. Pearson, & R. Barr (Eds.), *Handbook of reading research* (Vol. III), pp. 743–770. Mahwah, NJ: Erlbaum.

Lewis, C., Enciso, P., & Moje, E. B. (Eds.). (2007). *Reframing sociocultural research on literacy: Identity, agency, and power*. Mahwah, NJ: Lawrence Erlbaum Associates.

Manderino, M., & Castek, J. (2016). Digital literacies for disciplinary learning: A call to action. *Journal of Adolescent & Adult Literacy, 60*(1), 79–81.

Massey, D. (2005). *For space*. Los Angeles, CA: Sage Publishing.

Miller, S. M., & McVee, M. B. (Eds.). (2013). *Multimodal composing in classrooms: Learning and teaching for the digital world*. New York, NY: Routledge.

Mills, K. A. (2010). Shrek meets Vygotsky: Rethinking adolescents' multimodal literacy practices in schools. *Journal of Adolescent & Adult Literacy, 54*(1), 35–45.

Moje, E. B. (2007). Developing socially just subject-matter instruction: A review of the literature on disciplinary literacy teaching. *Review of Research in Education, 31*(1), 1–44.

Moje, E. B. (2015). Doing and teaching disciplinary literacy with adolescent learners: A social and cultural enterprise. *Harvard Educational Review, 85*(2), 254–278.

Moje, E. B., Luke, A., Davies, B. & Street, B. (2009). Literacy and identity: Examining the metaphors in history and contemporary research. *Reading Research Quarterly, 44*(4), 415–437.

NGSS Lead States. 2013. *Next generation science standards: For states, by states*. Washington, DC: The National Academies Press.

Peppler, K. A., & Kafai, Y. B. (2007). From SuperGoo to Scratch: Exploring creative digital media production in informal learning. *Learning, Media and Technology, 32*(2), 149–166.

Quinn, H., & Bell, P. (2013). How designing, making, and playing relate to the learning goals of K-12 science education. In M. Honey & D. Kanter (Eds.), *Design, make, play: Growing the next generation of STEM innovators* (pp. 17–33). New York, NY: Routledge.

Rogoff, B. (1995). Observing sociocultural activity on three planes: Participatory appropriation, guided participation, and apprenticeship. In J.V. Wertsch, P. del Rio, & A. Alvarez (Eds.), *Sociocultural studies of mind* (pp. 139–164). Cambridge, UK: Cambridge University Press.

Rowsell, J. (2013). *Working with multimodality: Rethinking literacy in a digital age*. New York, NY: Routledge.

Scribner, S., & Cole, M. (1981). Unpackaging literacy. *Writing: The Nature, Development, and Teaching of Written Communication, 1*, 71–87.

Shanahan, T., & Shanahan, C. (2008). Teaching disciplinary literacy to adolescents: Rethinking content-area literacy. *Harvard Educational Review, 78*(1), 40–59.

Street, B. V. (2003). What's "new" in New Literacy Studies? Critical approaches to literacy in theory and practice. *Current Issues in Comparative Education, 5*(2), 77–91.

Street, B. V. (2005). At last: Recent applications of new literacy studies in educational contexts. *Research in the Teaching of English, 39*(4), 417–423.

Tucker-Raymond, E., Gravel, B., Wagh, A., & Wilson, N. (2016). Making it social: Considering the purpose of literacies to support participation in making and engineering. *Journal of Adolescent and Adult Literacy, 60*, 207–211.
Vygotsky, L. S. (1978). *Mind in society: The development of higher psychological processes.* Cambridge, MA: Harvard University Press.
Wilson-Lopez, A., Mejia, J. A., Hasbún, I. M., & Kasun, G. S. (2016). Latina/o adolescents' funds of knowledge related to engineering. *Journal of Engineering Education, 105*(2), 278–311.
Wineburg, S. (2001). *Historical thinking and other unnatural acts: Charting the future of teaching the past.* Philadelphia: Temple University Press.
Winn, M. T. (2010). "Betwixt and between": Literacy, liminality, and the celling of Black girls. *Race Ethnicity and Education, 13*(4), 425–447.

2

LEARNING IN A PROFESSIONAL CONTEXT

LMNPO Case Study

Focus Literacies in This Chapter

Identifying, organizing, and integrating information: *Searching for information on the Web, using reference guides, material specification sheets, and product manuals*

Communicating with others for help and feedback: *Help includes providing or seeking assistance to complete a task or series of tasks that one could not do alone. More solicited, more hands-on, more involvement than feedback. Feedback includes providing or seeking advice, opinion, or consultation. Includes general guidance. Can be unsolicited.*

In part, learning by making has received widespread endorsements from technology leaders and politicians because it claims to support a production economy spurred on by innovation (e.g., Hatch, 2014; White House, 2014). That is, those people who can invent new products, or improve the old ones, will reap the benefits of the market and drive production (and money-making) forward as a whole (Schumpeter, 2003). This chapter describes a professional design firm, LMNPO, influenced by the maker movement and hopeful that they could be a backbone of the innovation economy by creating small-scale production capabilities for entrepreneurs. The fast pace of technological advancement and the shifting grounds of business mean that learning on the job is more important than ever (Manuti, Pastore, Scardigno, Giancaspro, & Morciano, 2015). As such, growth and development of the business is dependent on the growth and development of its employees. Work and learning are inextricable, especially at LMNPO.

In this case study, we are interested in the ways employees were expected to learn, why they were expected to learn in those ways, and the literacy practices that were related to those learning opportunities. Prominent literacy practices included identifying, organizing, and integrating information and communicating with others for help and feedback.

Learning in the Workplace

Workplace settings across the board require complex cognitive and social skills for people to be successful (Rose, 2005; Schon, 1987). Established companies often have communities of practice that reproduce both official and unofficial practices that are governed by long histories of rules, texts, and divisions of labor (Wenger, 1999). Researchers have conceptualized workplace learning along a continuum from informally picking up cultural practices through silent observation to more formal training programs or courses that may be completely didactic (Jacobs & Park, 2009). However, many companies are not structured for learning at all and often rely on informal learning opportunities (Eraut, 2004).

Most of the workplace learning we discuss in this chapter is intentional self-directed learning for building skills. Manuti and colleagues (2015) explain that in workplaces, "Informal learning recognizes that the acquisition of knowledge and skills in the work setting does not occur from organized programmes alone. Indeed, learning also occurs during critical moments of need embedded in the context of practice" (p. 5). The kinds of learning at DGF might be explained as a hybrid of formal programs, like trainings, and informal programs in which people learn just-in-time, as needed, or when the opportunity arises. As in many other workplace settings it means that workers have to figure out what they do and do not know. If they do not know something, they have to figure out how to learn it (Milligan, Littlejohn, & Margaryan, 2014).

Apprenticeships have been a structure for learning trades for thousands of years (Billett, 2016). Apprenticeships work by the apprentice engaging in tasks of the trade and slowly progressing to more complex or important tasks. Much of what we saw at LMNPO might be considered apprenticeships, but because the company was designed to be able to manufacture a range of products (listed later in the chapter) people were not apprenticed to any one trade or skill. Likewise, for much of the time they were told to figure things out on their own. There are now a number of websites that specialize in independent self-directed learning for workplace skills. They teach programming, finance, mathematics, software, design, and much more. Wikipedia is a trusted source of information for an introduction to how electricity works, or what a raster cut on a laser cutter is. Learners at LMNPO relied heavily on the Web to figure out solutions to the problems they were encountering. In what follows we describe the organization, the people who worked there, and tensions that arose between developing a staff that knew what it was doing for future customers' projects and serving current customer demands.

LMNPO Industries

LMNPO Industries (a pseudonym) was a small design engineering and rapid prototyping firm founded in 2013. It closed its doors in spring of 2015, during our study. L, M, and N are the surname initials of the three founders, all of whom were working in Artisan's Asylum, another of the spaces in our study, teaching and running their own businesses. LMNPO was located in what was formerly an industrial center in Greater Boston. The neighborhood, historically a home to both residential dwellings and manufacturing industries is now a center for innovation. Beginning in the late 19th century, the area was occupied by a company that created brass and copper tubing. This company is credited with expanding the area and buildings into the complex in which LMNPO, the making space, and a number of other incubator labs and small businesses were located. Most notably, an envelope company occupied the complex until 2010. Since then the LMNPO space, and the space that it shares, has been rented to various small businesses. The former Boston-Maine railway, now a commuter train to northwestern exurbs, still runs behind the building where LMNPO operated.

Employees

We include a description of the employees and their division of labor because a) there were not that many of them and b) it is important to understanding why the company engaged in the hiring, training, and education activities that it did, to understand how learning happened. For instance, the five top and mid-level employees came from the community at Artisan's Asylum. As we will show in Chapter 3, Artisan's Asylum embodies a set of values and practices that encourage people to both "do-it-yourself" as well as "do-it-together." Much of the learning is driven by individuals. There is an emphasis on technical expertise but also the idea that any skill can be learned.

There were ten total employees, including the owners, who worked at LMNPO. These included two owners, Marvin and Owner (not consented in the study because he was hardly ever there); a shop manager, Paul; a client relations and marketing specialist, Ethan; a design engineer, Elijah; a fabrication team consisting of Tim, Noelle, John, and Jason; and a lead programmer, Adam. Adam did not work for LMNPO, but a sister company also owned by the LMNPO owners called Bonzai (a pseudonym). Noelle was the only woman and everyone was white except Adam, the lead programmer, who was African American. The maker movement, and the engineering field in general, has been criticized for being largely populated by white males. LMNPO was no different. Every member of LMNPO was in some way previously connected to the owners or to the

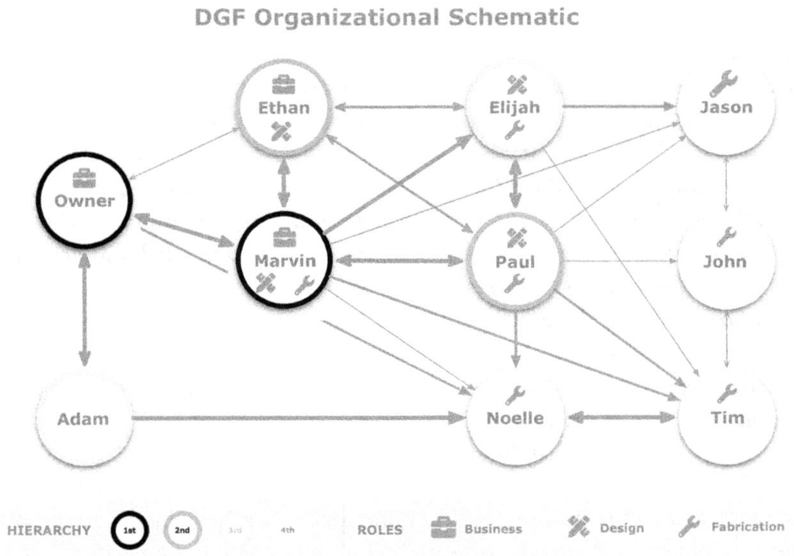

FIGURE 2.1 Organizational structure at LMNPO.

making space. The company itself began in the nearby member-run making space Artisan's Asylum, described in Chapter 3.

In Figure 2.1 we show the organizational structure of LMNPO, who talked to whom and at what intensities, their roles, and the dominant direction of the flow of information.

Owner, CEO

Marvin was the hands-on owner and manager of LMNPO. From a young age Marvin delved into the world of making, learning welding, apprenticing with a blacksmith, discovering programming, and dabbling in metalworking. He was also a journeyman silversmith and had a rich history of part-time and full-time work in information technologies (developing the information systems architecture for a publishing company), and working with his hands building boats. Marvin had a partner in the business (and in Bonzai) who was not consented to be in the study and in fact we only witnessed this partner on two or three occasions. We refer to him in this manuscript as Owner only when the data would make no sense were he to be omitted.

Design and Middle Management Team

Paul was the shop manager. Paul's background in construction gave him the confidence he needed to pursue learning the new skills and techniques required for his position. Additionally, he had training in woodworking and at the Audubon Institute. Paul considered himself a lifelong learner and was interested in making from a young age, yearning to learn the ins and outs of how things worked. *Ethan*, LMNPO's Creative Director, had a background in video production and illustration. Ethan was observed facilitating design meetings, drawing representations of products on the board, and fielding input. He was a liaison with clients, and he was in charge of projects through the scoping phase. He occasionally provided help for fabrication workers. Ethan was outgoing and maintained a professional appearance and manner. *Elijah* was a lead design engineer. As a child Elijah told everyone he wanted to be an inventor, and his passion for learning and making led him through an undergraduate degree in industrial design, various design and consulting work, and employment at LMNPO. After college he moved to Boston specifically to explore Artisan's Asylum, where he met Marvin. While with the company he was given added duties of shop manager in addition to CAD/CAM work. His eye for detail and meticulous work approach were evident as he taught other workers to use the laser cutter, computer-controlled three-dimensional milling machinery, and design software. All of the aforementioned team were drawn from the nearby making space.

Fabrication Team

Tim came to LMNPO as an unpaid intern looking to make a change after six years in customer service. Tim had some experience at a technical college in Boston, where he studied for approximately a year and a half. He began by learning the laser cutter, and then continued to guide himself to find and soak in new knowledge in this applied atmosphere. Tim was observed working on numerous projects in fabrication and electronics, and became very knowledgeable in the area of soldering. *Noelle* came to LMNPO with a background in early childhood education, nannying, and a passion for the visual arts. She used photography as a gateway to diving into other media. She worked for Owner's family helping to take care of his daughter, who is on the autism spectrum. LMNPO then hired her as a consultant on a project that addressed the needs of learners with autism. Working in fabrication, electronics, and coding at LMNPO allowed her to constantly learn new things and interact with new materials. She described herself as a generalist. Noelle was a self-described logical thinker who enjoyed coding at LMNPO, and gained confidence in her abilities to fix things at work and around the house. *Jason* spent most of his time at LMNPO learning the design tools, such as Autodesk design software and Fusion360. He held a graduate degree in film and continued to explore and create through film outside of work. *John* was involved in welding and fabrication at LMNPO, where he was hired after working as a maintenance man for the building. John did not want to be interviewed for the study. *Adam* was the lead programmer at LMNPO, but was primarily involved with Bonzai projects. He was not interviewed for the study, nor was he observed more than a few times.

Observing at LMNPO Technologies

We began observing at LMNPO in January of 2015 and finished when the company shut down in the middle of May. We observed a total of 34 times, averaging about two observations a week and totaling approximately 102 hours. We also conducted semi-structured formal interviews with seven of the ten employees. The interviews were strung across the five months of our time there. We recorded our observations of activities through fieldnotes, informal conversations and questioning, and photographs. Observations were guided by our developing literacy practices framework and were mainly focused on fabrication activities, but also included design, fabrication, and tinkering processes as participants made a variety of objects including, among others, the following:

- A large "touch easel" screen for young people with autism.
- Laser cut boxes.
- Interactive displays for a room at a museum.
- A small injection molding machine.
- Various machined parts.
- A specialized coffee cup for single pour coffee.
- A day's travel case for diabetic medicine and supplies.
- PCB boards for electronics products.
- A kinetic sculpture for a university making space.
- Electronic paired lovebird wristbands that when tapped by one partner, would be felt by a remote other.

We tried to observe design meetings when possible, however three factors limited our ability to do so: 1) LMNPO felt clients would be uneasy in our presence, 2) non-disclosure agreements prevented LMNPO from including us in many design meetings, 3) there was not much new work for LMNPO to do during our time there. Therefore, the kinds of activities we observed were mainly limited to fabrication activities.

Hiring and Training Practices

Marvin's own education and vocational history shaped much of the literacy and learning practices at LMNPO. At the same time, Marvin's experiences also mirror many of the values and experiences we have found to be touted by the maker movement such as distributed expertise, just-in-time learning, leveling up skills, open access to information, and following one's own interests and passions (Anderson, 2012; Hatch, 2014).

Marvin drove the hiring process, and he described his approach: "I've made it a point of hiring career transition people all through my time, because people educated in the formal topic I'm working within are blind to a lot of the aspects of it." Marvin hired motivated individuals whom he could teach on a just-in-time basis. Because they have little to no previous educational experience in engineering design or fabrication, he could shape them to work in his preferred manner and teach them skills when the need arose in their work. He hoped to find people he thought had "common sense," regardless of whether they walked in knowing how to use a tool or not. Paul explained Marvin's hiring practices as an investment, "Marvin [is] always more willing to invest in a staff member for the long term than just getting somebody maybe cheaper that will decide to move on in three months or whatever." Or, as Marvin explained, "So, these are people who in less than a year have gone from what would normally be minimum wage positions, really, to I am making money on them, to in another year, they will be actually very valuable people."

Marvin valued the diversity of perspectives among the staff and he invested in their future with the company. He took on the three floor-staff members as interns in apprentice-like positions. Initially, they worked for free or very little pay, and then their pay increased as they learned skills and continued to work for the company. Tim, Noelle, and Jason were all brought on board without formalized experience in fabrication and were being trained in various aspects of production until they were able to be self-sufficient.

It was common for Marvin to hire people within his network, those he felt had potential. There was tension surrounding this, however, as other members of the management team described in their interviews. For instance, Ethan stated, "I have argued for a moratorium on hiring people we know. I have raised my hand and said we should not hire a single person we know for anything." He explained that with the current rate of production, it was hard for staff members to take the time to train new employees. Ethan did not want to train new employees and reported that this put a strain on management as they spent too much time checking in with people about their training rather than completing work for clients. However, he noted that Tim started as an intern and "pick[ed] stuff up quick." Ethan explained that Tim learned from experience, as did Noelle.

> I think the approach, at the time, we took was right but now we've found ourselves in a position where we'll wait, we have in some cases and we'll end up wasting time is we have to factor training in versus execution. So, if people have to keep coming up to speed on things in order to get a project done, we basically can't bill a client for the week it's going to take us to get someone trained.

In Ethan's mind, there was a tension between making money immediately versus hiring someone that would take months to train. While Marvin's view was that learning adds long-term monetary value to the company in shaping an individual new to the field, Ethan (who was in a managerial role) identified the short-term pitfalls of all that attention to learning.

He discussed the nature of bringing in employees from different backgrounds, but posited that his main concern was client satisfaction: "from my perspective, once the client comes in, I want to see their thing get done right and get done on time." He elaborated:

> And if having someone in the shop pick up a new skill to get it done, if it doesn't impact either of those [done right and on time], I'm all for it, because it builds the skills of our team and you know, we don't have to bring someone new in. It's from a business perspective, it's a stronger way to approach the project—we use the team we have. We don't have to spend extra money. They get a skill and we still get the project done.

Despite the tension, and perhaps because there was not much paid work, floor staff spent a good deal of time learning to weld, run the laser cutter and 3D printer, and other tasks.

Learning at LMNPO

Learning on the Job

Employees of LMNPO were expected to learn the tools, processes, and skills necessary to complete their assigned tasks. This expectation was supported by more experienced employees, who would offer forms of help situated in the particular moments of making that employees encountered. What we observed was very similar to descriptions of apprentice type models of learning (Lave & Wenger, 2001). Learning organized around doing the work of the firm was supported by oral communications, short instructional conversations, and, at times, leaving the employees to figure some things out for themselves just by doing them; the backbone of learning at LMNPO was oral communication, demonstration, and learning by doing (see Figure 2.2).

Tim explained that he happened to learn well from verbal instruction, as he did when he learned the laser cutter upon first beginning at LMNPO. A more experienced employee guided him through the process using just verbal instruction and action. Tim elaborated,

> It was "hey here's the button where you can do this and here's the story about how if you leave this setting on while running the laser, it won't beep when it's over." So three hours later you wouldn't be wondering like, "Oh what happened to that cut." It was very verbal based which I respond well to.

Tim was expected to complete a task using the laser cutter, and he was given a place to start with some oral instructions coupled with a demonstration. It happened that

FIGURE 2.2 Paul, Tim, and Jason working on troubleshooting a large touch table console they were designing. Paul is leading Tim and Jason using a diagram he drew of the circuitry inside the console.

the oral, showing, and doing styles of learning and instruction worked well for Tim. We also watched Jason spend hours on his own, without much interaction with others, designing storage boxes on the CAD software to be laser cut with only occasional questions to Elijah. Jason echoed the notion of experiential learning,

> Yeah, I feel like that's an easier way for me to learn and perhaps for people to teach. Is just by, "Okay well this is it. This is how we do it I'm going to show you how to do it once then you're going to do it, and ask me any questions and then okay do it." And I feel like for both people who can learn through reading and are visual learners. I feel like that's a good way for both types of people. I don't think that there's any restrictions because you're actually doing it.

Learning how to make was situated within actually making things. There were no prerequisite special skills or knowledge. LMNPO employees could operate sophisticated software and machines simply by practicing with the tools themselves, and, after initial introduction, largely on their own. Often this learning-by-doing produced specific, contextualized problems with which to learn. A large part of the learning on the job occurred around the posing and solving of problems when employees communicated with one another to ask for help. Jason explained that he knew he needed the assistance of someone more experienced when he ran into roadblocks in his learning. By communicating with a more knowledgeable staff member he was able to work his way through any problems.

> I know that at some point I'm going to hit a roadblock. And I'm going to need to say bring in somebody who's more experienced, and be like can you help me with this? Because I'm stuck here, and one of the great things about everyone that works here is that it's usually a communication of "Ok, well, why are you stuck? Or why do you think you're stuck? And how can we work through this? What do you think you need to do to get success from this thing?" . . . So the intuitive mixed with the conversations I guess with the Socratic teaching is really the best way . . . one of the best things that I've experienced here in terms of that Socratic communication.

Jason's experience was typical within the learner-oriented LMNPO. He was mostly left to learn on his own, but when he needed help, more experienced members of the group were ready to provide it, albeit in rather specific ways. He highlighted the Socratic style of instruction in which the more experienced employees asked the newer workers how they would solve the problems. Newer employees learning tools were not told directly how to solve problems, or how to fix the issue they confronted, rather they were asked what they thought and given space to explore their own ways of understanding the problems. They were expected to find the information they needed on their own, even if that information came through oral conversations with more experienced colleagues. The critical skill they did need, not tied to any one tool, was to have the confidence to solve problems on their own.

There were other kinds of learning at LMNPO as well. For instance, Tim and Noelle were asked to learn how to fabricate printed circuit boards for a company, Alternative Manufacturing Inc. (AMI), that was subcontracting to LMNPO. They wanted to take on learning new skills because that was part of the reason they had joined LMNPO in the first place. Tim and Noelle both expressed interest in continuing to build upon the skills they learned through their work with AMI. The pair worked together making circuit boards, a process they learned during an intensive, hands on training in Maine towards the beginning of their employment with LMNPO. Noelle explained that she first picked up a soldering iron to heat up an insert, not to actually solder something. This initial interaction with the tool was enough to spark a deep interest in the process, something Marvin picked up on. "He also taught me how to pull solder and how to connect wires and disconnect wires. And I fell in love with just the visualization of seeing the melted solder. That was the coolest thing." Through her expressed interest in soldering and putting together electronic components, Noelle was given the opportunity to follow her interests and formally train with AMI (see Figure 2.3).

Tim and Noelle were in constant contact with a representative from AMI as they worked. Communication between them occurred when the two spotted an error in their spec sheet, or when they encountered a problem. Thus, learning happened continually and on the spot as they worked. The pair was in a unique position as they worked independently of LMNPO and the expertise of its management team. They also communicated verbally while working together, verifying each other's work and teaching each other. In one instance, Noelle asked Tim to teach her to load a specific cartridge into a machine. The two had just solved a problem with another machine, communicating verbally and working together to adjust the machine. In lieu of explaining the process to her, Tim simply stood back and watched as Noelle successfully attempted to load it. He answered her questions with a simple "yes," allowing her to learn experientially and to solve problems on her own, much in the same way they had been taught by the more experienced employees.

Tensions of Learning at LMNPO

Within this situated approach to learning, not all learning was automatic. The "show once and step back" approach did not work well for everyone, especially as the company needed to complete fabrication of professional-level products. There were tensions within and inherent to the learning arrangements at LMNPO that included what employees were supposed to learn, when they were supposed to learn it, how they were supposed to learn it, and how that learning potentially impinged on customer satisfaction and cost effectiveness.

Most, if not all members of the LMNPO staff wore many hats. Tim worked in fabrication and with AMI. Paul managed the shop, taught employees new skills, and helped with fabrication. Elijah designed and helped to make products. Marvin was involved with everything, including bookkeeping and meeting with potential and current clients. LMNPO relied on the versatility of its employees, which also led to tension around the what and when of employee learning on the job.

Noelle came to LMNPO through a connection with Owner, who brought her onboard as a consultant to the Bonzai large touch table project for children with autism. This soon led to her learning code with one of the founders

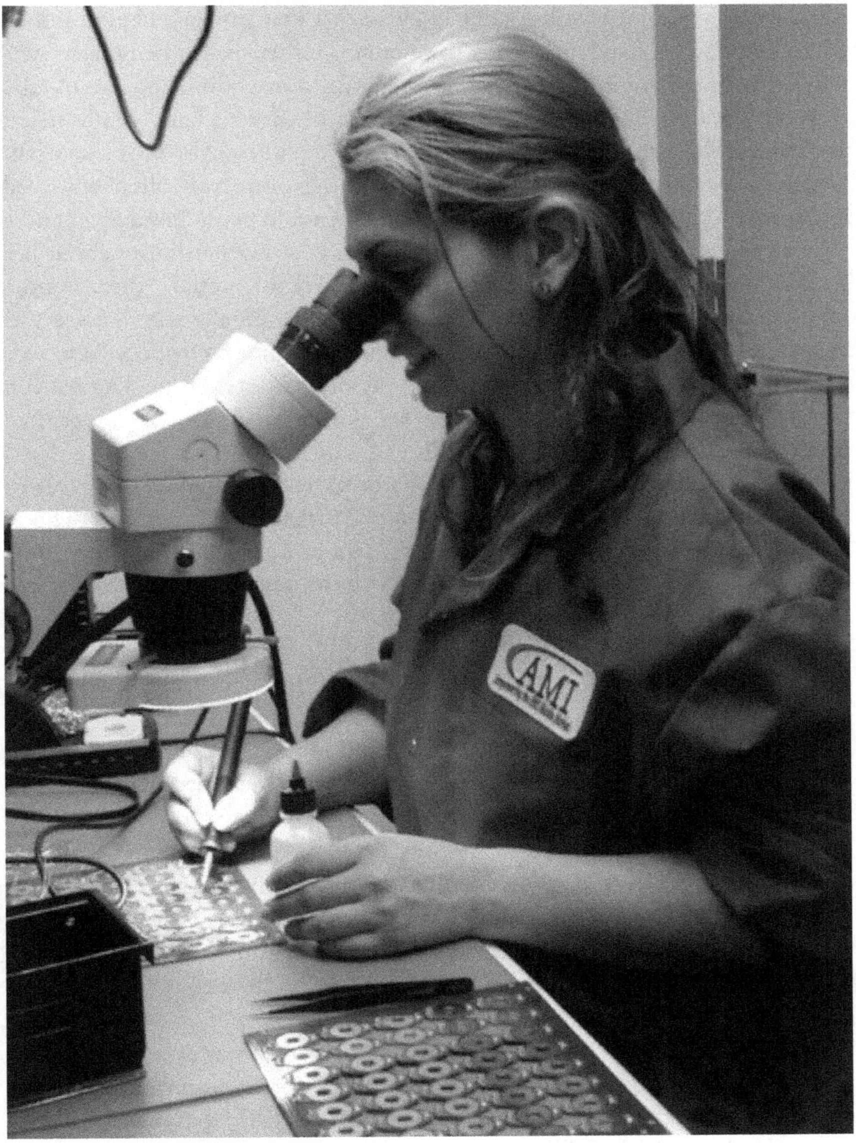

FIGURE 2.3 Noelle soldering, one of her many skills.

and eventually with Adam who was the main programmer for Bonzai. Soon after, Marvin encouraged her to learn soldering, and she attended the AMI training in Maine. As such, Noelle's work was spread across multiple domains. She described the challenge of very quickly learning to code, "It was seriously jumping. Every month I was jumping to a different language, which was a lot. It was a lot of stimuli but I'm glad that that happened because now when I sit down I have something to reference rather than it being all new." After this initial training Noelle took a hiatus from coding and began fabrication work for LMNPO, during which she learned to sand, grind metal, and solder. She also learned the basics of the laser cutter. Also eager to learn, Noelle constantly sought to pick up new things, and was often observed learning multiple concepts simultaneously. She noted being hired as a generalist, expected to perform multiple roles. This kept her passionate about her work and allowed her to fulfill her desire to learn. She noted, "there's nothing here that I don't want to learn." The tension for Noelle was that she was expected to learn multiple skills and found herself jumping from one practice to another without getting a good grasp on any particular one. The tension for the company was that she was not yet skilled in any one area, yet they needed her to be multidisciplinary.

Another tension at LMNPO existed around self-directed learning versus expert-led learning. For example, Noelle noted that she did a great deal of book learning and self-directed learning when she first came to LMNPO. After some time, Adam began teaching her to code, which she described as much different than learning from books.

> And now working with Adam it's really different because he's actually teaching me versus learning from books. And it's much more easy to relate and easy to understand because of both the fact that I've learned a lot of it previously and he's giving me in-person examples that I can relate to and it makes much more sense.

In an example of identifying, organizing, and integrating information, she was better able to process this information, and piece everything together from her previous learning by talking with someone and working through real examples. Again, the apprenticeship model seemed to be the best way to learn at LMNPO.

In his interview, Tim explained that the ability to learn was a trait that resulted in value as a LMNPO employee, because Marvin often directed employees to sources and expected independent learning. LMNPO epitomized learner-centered values, and yet, this alone was not enough to resolve tensions in how people learned. Not having proper guidance as one learns proved at times to be detrimental. In one instance Tim was tasked with filling holes on a wooden frame with Bond-o, a material he had no prior experience with. Lack of supervision from employees experienced with the material and pressure to complete the task combined led to Tim botching his attempt. He noted that it would have been better to learn how to use the material before they were in the middle of creating a professional-level product. When describing the project, he stated,

> I screwed up pretty badly and ended up making us spend another two or three hours sanding one of the touch easels because I didn't use Bond-o correctly. But the only person in the shop who was qualified to watch over me and verify that I was doing it right was not present. So it's one of those cases where learning in the middle of a project can really be detrimental especially if you don't have guidance. So being frontloaded on the knowledge probably would have been way better.

In this case, the apprenticeship model broke down because Tim, the apprentice, was left alone before he was ready.

In some instances, Marvin was observed modeling various fabrication techniques. This was often done when the company was under pressure to complete a project for a deadline, or when Marvin seemed frustrated himself. On one occasion he demonstrated how to more efficiently use a hammer for John (Fieldnotes, 5/1). On another he took over soldering a part from Tim, as he could complete this task more accurately and efficiently (Fieldnotes, 3/11). On both occasions he was observed narrating his actions and offering instructions. Tim's sentiments echo the notion that this would have been helpful if it had occurred more often and employees were not left in the dark to work independently. Yet everyone was trying to complete multiple tasks and these learning arrangements may have required more resources than the small firm could provide.

Because learning relied so heavily on oral communication, demonstration, and self-directed problem solving, a lack of communication regarding employee logistics created tension among employees. Tim explained that he often came into LMNPO with no idea what he would be working on that day. He said that he never knew if he would get to do something interesting, learn something new, or do "grunt work" for hours. It happened during the day too. For instance, on one occasion Tim was observed being directed to abruptly switch tasks from searching for a tool to fixing a computer, "Of course at the end of lunch they're like we need you to set this thing up for us" he quipped, setting a keyboard and mouse down on a worktable.

For employees, there were also tensions between the iterative, experiential, and the on-the-job learning process of LMNPO employees and management's need to ensure quality control and customer satisfaction. Ethan noted that though Tim and Noelle were fast learners, time was wasted on their training when that time could be spent fabricating. Essentially, as employees learned what they needed to in order to finish a project, LMNPO lost money and time. As the client liaison, Ethan walked the line of wanting to see others learn and wanting to efficiently make the best possible product. He explained his view on Marvin's hiring practices in relation to this challenge:

> I think the approach, at the time, we took was right but now we've found ourselves in a position where we'll wait, we have in some cases, and we'll

> end up wasting time is we have to factor training in versus execution. So, if people have to keep coming up to speed on things in order to get a project done, we basically can't bill a client for the week it's going to take us to get someone trained in a way to work on something. So you're either losing a week's worth of time that could be billable work to train someone to do something.

Ethan also explained that employees should be given time out from current projects to learn a skill.

> If we want Noelle to have the skill, let's put time on the calendar for her to do like a test run so that she's not feeling the pressure of just trying to get a project done. We're not feeling the pressure of the timeline for the project.

Pressures to learn and to perform placed undue stress on employees. Elijah echoed that, noting that if possible, the team sought to give each person the time to learn something as best as possible. This was the case for Jason, as he was able to gain experience learning on smaller projects, such as crafting cartridge holders for the 3D printer. This project allowed him to learn design software and the laser cutter through iteration without worrying about meeting client deadlines.

Impacts of Learning Beyond LMNPO

According to Jason, having an inherent interest, motivation, and desire for hands on learning is essential to learning. He noted, "wanting to have that mental wonderlust of I wonder what this does, like, how can I make something better?" This "wonderlust" is what drove individuals to join LMNPO's ranks, and ultimately to continue learning outside of its walls. Many LMNPO employees told researchers that they took an interest in learning in their personal time.

A theme of personal interest and motivation was revealed as Paul and Eric explained taking personal time to learn information. Paul took the time in his personal life to learn information and to comb through information from a variety of online sources. Eric brought up a similar theme as he described watching YouTube videos and observing a seasoned professional to learn welding. Observations were a large part of learning for those at LMNPO. Ethan took a reverse engineering approach, looking at something and wondering how he could make it himself. He then watched videos and sought out knowledgeable individuals to help him. As such, observation and conversations were a large part of his learning on a personal level.

Tim came on board with a strong desire to learn and noted the value in gaining useful skills that in addition to their potential compensatory value.

Tim told a researcher that his work at LMNPO informed his personal making as he fixed his friends' laptops and hoped to implement new skills such as soldering. Tim was also eager to complete outside homework like projects assigned by Marvin, such as learning about polarities, or how touch screen components work. In these instances, Tim would come to Marvin with a question, who would then direct him to a website or online search. He would also pose a problems or challenges for Tim to solve. Tim would return in the following days with newly acquired information to hash through with Marvin.

Clearly, the employees at LMNPO had much to learn. They also *wanted* to learn, which was a large part of why they worked there in the first place. However, given LMNPO's status as a start-up without much capital, the relative obscurity of the tools they were using, and the rather complex products that the business was producing, there were tensions between employee learning and producing quality products. As a maker-centered workplace though, there are lessons to be learned for educators including those looking to prepare young people for the workforce.

Learning in a Making-Centered Workplace

The maker movement has emphasized and valued the *networked* and *open-sourced* characteristics of information and workplaces that are learning-oriented (Hatch, 2014). LMNPO also subscribed to these values. LMNPO was an inherently interdisciplinary company. Employees needed to know how to do many tasks and work with different tools, just as one would at a making space. Learning in workplaces is no longer confined to the rules, manuals, and proprietary information of the company. Employees are expected to learn from open-sourced and networked information. And at LMNPO they were also expected to play, to learn on their own as if in a sandbox.

People were asked to solve problems and when they ran into trouble, there were different ways of supporting them. Some support was through conversation, some was through physically showing someone what to do. The company needed to provide multiple kinds of learning interactions, some more guided than others, to meet employees' needs and simultaneously meet the needs of clients.

Learning within a maker ethos constructs the learner as self-motivated and self-sufficient, like an entrepreneur would be. At LMNPO, we found that to be the case as well. Yes, other people helped, but the main person responsible for learning was the individual learner, not necessarily management. It is a different approach to responsibility than in most school experiences. It has implications for learning in making because it asks learners to know how to find information on their own and how to talk with others to get help.

Formal institutions of learning have many prerequisites for advanced courses. For instance, one needs to have taken Algebra to take Geometry in most places.

Even when courses are advanced, curriculum cycles often assume that students need to know how to do one kind of operation before they do another. Observations of learning at LMNPO suggest that learning while making does not require previous knowledge. Much of what needs to be learned can be done just-in-time. What one does need to know how to do is identify problems and search for information. Searching for information, as we will show again in Chapter 4, is a skill that involves knowing where to go and who to ask.

The need for versatility created tensions that many educators will recognize. As in the case with Noelle, much of the learning was about breadth, not depth, of knowledge. Employees needed to know enough to operate several tools, but they did not need to become masterful. They needed to be able to transfer understandings of how tools worked from one to another. In making spaces, learners and educators might feel overwhelmed by the amount of information they need to know to accomplish tasks. They can be buoyed by two facts: 1) if you don't know how to do something it is okay because someone else does and they are usually within reach, and 2) one does not need to be a master craftsperson to create, one only needs the desire. At the same time, making spaces need to be able to provide the structures for learning. These can include a range of learning opportunities: more formal classes, apprenticeships, and self-directed learning of the sort found at LMNPO among them.

References

Anderson, C. (2012). *Makers: The new industrial revolution*. New York, NY: Crown Publishing.

Billett, S. (2016). Apprenticeship as a mode of learning and model of education. *Education + Training, 58*(6), 613–628.

Eraut, M. (2004). Informal learning in the workplace. *Studies in Continuing Education, 26*(2), 247–273. doi:10.1080/158037042000225245

Hatch, M. (2014). *The maker movement manifesto: Rules for innovation in the new world of crafters, hackers, and tinkerers*. New York, NY: McGraw Hill.

Jacobs, R. L., & Park, Y. (2009). A proposed conceptual framework of workplace learning: Implications for theory development and research in human resource development. *Human Resource Development Review, 8*(2), 133–150.

Lave, J., & Wenger, E. (1991). *Situated learning: Legitimate peripheral participation*. Cambridge: Cambridge University Press.

Manuti, A., Pastore, S., Scardigno, A. F., Giancaspro, M. L., & Morciano, D. (2015). Formal and informal learning in the workplace: A research review. *International Journal of Training and Development, 19*(1), 1–17.

Milligan, C., Littlejohn, A., & Margaryan, A. (2014). Workplace learning in informal networks. *Journal of Interactive Media in Education, 1*(6).

Rose, M. (2005). *The mind at work: Valuing the intelligence of the American worker*. New York, NY: Penguin.

Schön, D. A. (1987). *Educating the reflective practitioner* (p. 27). San Francisco, CA: Jossey-Bass.

Schumpeter, J. A. (2003). *Capitalism, socialism, and democracy*. New York, NY: Routledge.

Wenger, E. (1999). *Communities of practice: Learning, meaning, and identity*. Cambridge: Cambridge University Press.

White House Office of the Press Secretary. (2014). Retrieved from https://obamawhitehouse.archives.gov/the-press-office/2014/06/17/presidential-proclamation-national-day-making-2014

3

SPACE MEDIATES COMMUNICATION

Artisan's Asylum Case Study

Focus Literacy in This Chapter

Communicating with others for help and feedback: *Communication involves fundamental literacies of speaking and listening. In the STEMLiMS framework it means the ways in which people ask for and give help, the ways in which they receive and offer feedback, and the ways in which they build knowledge together in face-to-face interactions.*

This chapter presents a case study of the Artisan's Asylum, an adult making space. In creating and sharing the case, we discuss how the physical, cultural, and social spaces of Artisan's Asylum produce social relations and mediate the ways in which people communicate with and learn from one another as they work on individual and collaborative projects (Lefebvre, 1991; Tucker-Raymond, Gravel, & Pierce, in press).

Face-to-face communication is not a straightforward process of information transmission, send and receive. It is mediated by complex social rules, interlocutors' identities, and as we argue in this chapter, the arrangement of physical, cultural, and social spaces. How space is arranged in making spaces is integral to cultures of making and can contribute to makers' and learners' investments in networks of expertise. Those networks of expertise are important because they are where people learn new information and new ways of doing things. Knowing how to communicate, with whom, and when, aids makers in accomplishing their work.

Human beings or societies produce space together to structure, open up, and limit activity (Lefebvre, 1991). The spaces that humans inhabit have certain value structures and ideologies because they are constructed by those who live in them with their own values, ideologies, and ways of being in the world. As human beings, we live life participating in the spaces that we construct; we are both in the space and responsible for its existence. The building that the Asylum occupies does not constitute the space. At one point it was a factory, at another, a warehouse. It is a space because of the people in it and what they do. It is the product of people's interactions with each other around different forms of activity (Massey, 2005). This means that the ways in which people communicate and what they are doing when they communicate create the space. The space, in turn, mediates the communication. As new members join, bringing with them their own histories, values, and practices to the Asylum, the social, cultural, and physical spaces are incrementally, yet continually, reconstructed.

Relationships and connections between multiple people across spaces have been characterized by sociologists and others as networks (Lin, 1999). In this chapter, we also characterize relationships between people as a network. However, we think of networks as dynamic, and people not as static nodes, but as moving through space and time. It is what others have called meshworks (Lefebvre, 1991; Ingold, 2011). Thinking of networks dynamically, or as meshworks, contributes to our exploration of how individuals come together to construct the Asylum space and inversely, how that space is constructed through their communications with one another.

Artisan's Asylum

Artisan's Asylum—the name was decided on through a poll on the Web, including the apostrophe before the "s"—describes themselves as dedicated to supporting and promoting "the teaching, learning and practice of fabrication" (https://artisansasylum.com/about/). The "Asylum" or "Artisan's" (we will refer to the space in both ways, as members do) is governed by an elected board of members, and a skeleton staff hired to handle administrative issues (e.g., insurance, utility contracts, managing memberships, and access). Members have opportunities (in person and through email listservs) to voice concerns, share suggestions, and to monitor and react to incidents and events in the space. Everyone takes out the trash. With the distribution of responsibility, members need to cooperate with each other in ways that also maintain individual productivity so that it is worth it to stay. This constructs individuals in relation to each other as mutually dependent, and we argue it creates opportunities for learning.

Making spaces "share some aspects of the shop class, home economics class, the art studio, and science labs. In effect, a maker space is a physical mash-up of different places that allows makers and projects to integrate different kinds of skills" (Dougherty, 2013, p. 9). The Asylum, currently in its third location, is situated in three large, connected warehouses that were once part of an old envelope factory. It is well-resourced with tools that have been donated or that community members have purchased by pooling their resources.

The physical size and layout of Artisan's Asylum is central to the ways in which work gets done, how people talk to each other, and how they collaborate. For the most part, there are "shops" that house the communal tools and workspaces and then there are "studios" that individual members or groups rent from the organization. There is also a small office for the executive team and a common space for people to meet, eat lunch, and chat or work. See Figure 3.1 for a map of the space at the time of our study.

The map in Figure 3.1 shows the numbered studio spaces as well as the shops marked by name. Multipurpose Room 1 houses the computers used for CAD design and other tasks. Multipurpose Room 2 was a meeting space with tables. Sometimes meetups or classes were held in Multipurpose Room 2 that did not fit in one of the shops, for instance, leatherworking. Most of our focal participants' studios were in building 10 and a couple resided in building 8. The densest activity at Artisan's Asylum emanated from the social area and multipurpose room outward. The other shop that we observed with the most activity was the woodworking shop; however, all spaces were well used. See Table 3.1 for a complete list of shops and some of their tools. Names of shops in the table and on the map do not completely match. For instance, the table is labeled Digital Fabrication and Design Lab and the map shows "laser cutter shop." In reality, the shared 3D printers are also in the "laser cutter shop." The metal casting shop is not labeled at all, although it is in one of the blank spaces near studio numbers 65 and 66. Both sources of information were pulled from the website. The inconsistency is perhaps a result of the growth of the Asylum and the nature of member-run spaces in which people volunteer to accomplish different tasks.

Shops are voluntarily maintained by members. Everyone is responsible for cleaning the space (e.g., taking out the trash, sweeping sawdust) and Shop Stewards oversee shop and tool maintenance. Members pass a tool training to gain privileges for use. Most times members have to fix the tools when they break down. The diversity of shops leads to many different disciplinary practices, traditions, and ways of being and knowing. Makers often work across different shops, bringing processes, materials, and practices from one space to another. Artists, designers, and engineers also bring different yet overlapping ways of thinking about the world. These overlaps contribute to creativity in manufacturing. Many members are proficient with the tools in more than one shop.

Studio spaces for rent can be as small as a shelf for storage or 400 square feet shared by a collective of people. Most people who rented had 50–100 square feet of space for their own studio which they populate with their own projects, personal tools, and materials. Every studio space was open on one side. While this did sometimes lead to conflict, the Asylum was not often open to the public and people did not worry much about having their items stolen.

The whole of the organization was run, like the shops, by members. The official organization included an elected board of directors and two paid positions (the executive director at the time of our study did not take a salary, the

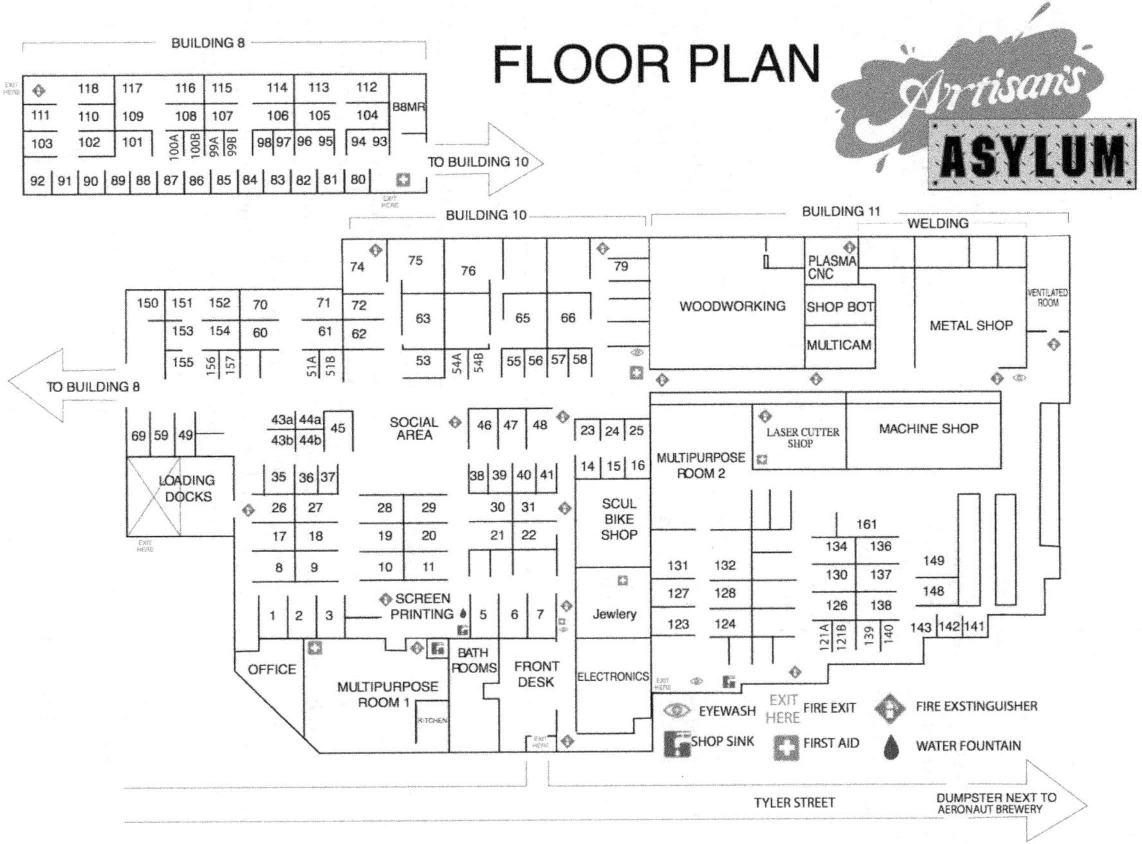

FIGURE 3.1 Map of Artisan's Asylum showing shops and studio spaces.

TABLE 3.1 Shops and examples of tools

Bike Shop base of operations for bike building, repair, and bike gang	**Computer Numeric Control Router Shop (CNC)** plasma cutter, Shopbot, multicam	**Digital Fabrication and Design Lab** 3D printers, laser cutters, vinyl cutter	**Electronics and Robotics** soldering stations, magnifiers, stereo microscope, oscilloscopes, hand tools
Fiber Arts sewing machines, overlock machine, iron	**Jewelry and Metalsmithing** six jewelry benches with torches, flex shaft, annealing pan and bench pin; buffing station; arc welder	**Machine Shop** mills, lathes, drill presses, horizontal and vertical band saws	**Metal Casting** kiln and crucibles
Prop Shop airbrushing booth, foam cutting jigsaw, sanders, casting table	**Screen Printing Shop** four color press, one color press, darkroom, drying racks	**Welding Shop** TIG, MIG, spot welders, hand held plasma cutter, oxy-acetylene, sand blasting cabinet	**Wood Shop** table saws, jointer, planer, drill presses, wood lathe, mitre saw, band saw, hand held tools

director of education and the director of membership were half-time only and the building manager was full-time).

There were tuition-based classes, almost all taught by members. Topics were offered in all of the shops. Some topics, like fiber arts or leatherworking, were not embodied by shops. Classes were open to those 18 and over. Most beginner classes invited students to make with accessible projects that were embedded with the skills necessary to pass a tool training and use a tool independently. For instance, through learning to make a wooden box, one becomes eligible to take tests for independent use of the table saw, mitre saw, planer, jointer, and hand tools. That class occurs over several weeks. Some classes take place in one evening. Teachers received 50% of all tuition and many used that money to cover the costs of their monthly space rent. At the time of this writing there were classes in marketing and business development, and, during our time there, an arts-marketing specialist rented space in-house.

Participants

We interviewed and shadowed 20 members of the Asylum in the six months that we were there. We selected a smaller sample of members due to constraints of time and availability, but also because we wanted rich data. Most of our participants were long-time, full-time, day-time members. We interviewed those who called themselves artists and those who thought of themselves as engineers. Others were craftspeople, like Matt, a former lawyer who made furniture from reclaimed wood. In all, we worked with jewelers, sculptors, woodworkers, visual artists, a hardware developer, a mechanical engineer, an electrical engineer, a software engineer, hobbyists working in electronics, a leatherworker, a welder/artist, and staff members. All participants in Table 3.2 were interviewed. There were other participants, not formally interviewed, that are not included in the table. All of our participants were white. Camila had recently moved to the US from Brazil. Everyone else, to our knowledge, was born in the United States. In Table 3.2 we list the participants' names, their main making activities and generally when they were at Artisan's Asylum. Names of participants are the names they gave us (some are not their birth names). All others are pseudonyms.

Our methods captured a representative sample of the types of making identities people used to describe themselves and gender. Our sample is not systematic in that we did not attempt to capture a representative population of those who used or even had space at Artisan's, which includes many more part-time makers and class takers. Rather, we focused on including a variety of perspectives from those who could provide the most information.

Network of Communication

As part of our research into how people communicated, we asked our participants to name everyone they could think of who they went to for help, feedback, or collaboration, and who came to them. This resulted in a network of members and the people with whom they interact. Figure 3.2 is a representation of that network. Lines represent those connections of communication. Icons refer to broad maker identity categories (e.g., engineer, artist). The size of the icon refers to how many references that person received (e.g., Chris T. and Brandon were mentioned by other people the most).

In the visualization of the network, 13 engineers and 9 artists. All of the engineers are men and all of the artists women. However, the three people who identify as engineers/scientists *and* artists are also women (Ecco, Camila, Carolyn). And the artist/craftspersons are a man and a woman. While this does represent a somewhat gendered division of how individuals identify, there is room for hybridity, and from a population standpoint, Artisan's is relatively gender-balanced, especially considering the national and commercial narrative of making as a male dominated activity (Buechley, 2013; Chachra, 2015). In part then, this gender balance is due to an expansive view of making as more than electronics and robotics. However, the narrative that making is an activity that white people do (Vossoughi, Hooper, & Escudé, 2016) is reinforced

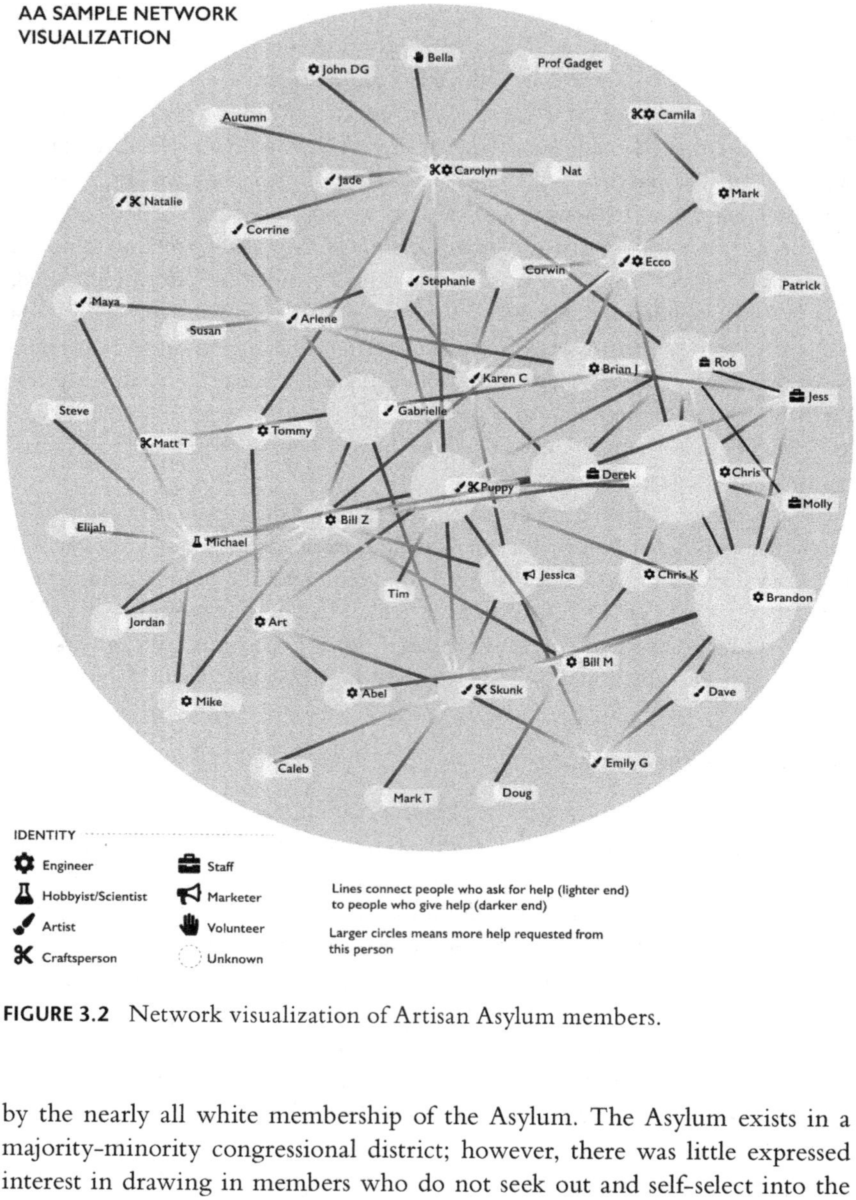

FIGURE 3.2 Network visualization of Artisan Asylum members.

by the nearly all white membership of the Asylum. The Asylum exists in a majority-minority congressional district; however, there was little expressed interest in drawing in members who do not seek out and self-select into the Asylum. Although the Asylum participated in community events, such as the annual Marshmallow Fluff Festival, and Artists' Open Studios, there were no organized efforts to recruit people of color. Due to low staffing, such efforts would likely have to be member-initiated.

Despite its racial homogeneity, transdisciplinarity, and hybridity were built into the Asylum from the beginning. The organization was started by a couple who wanted to share tools with their friends. She sewed, he built robots. From

TABLE 3.2 Focal participants, main making activity, and intensity of time at Artisan's Asylum

Participants	*Making Activity*	*Time Spent @ AA*
Arlene	Jewelry	Full Time/Day Time
Art	Electronics and Mechanical	Part Time/Night Time
Bill	Electronics and Robotics	Full Time/Day/Night Time
William	Software Design, *Board Member*	Full Time/Day Time
Brandon	Computer Hardware/Software	Full Time/Day Time
Camila	Architectural Design	Full Time/Day Time
Carolyn	Woodworking, Fiber Arts	Teaching/Former Member
Chris	3D printing, Engineering	Full Time/Day Time
Derek	*Executive Director*, Lasers	Full Time/Day Time
Ecco	Fabrication, 3D & Visual Art	Full Time/Day/Night Time
Emily	Visual Art/Map Making, *Board*	Part Time/Day Time
Jess	*Member*	Part Time/Day Time
Karen	*Director of Membership*, Dance	Full Time/Day Time
Matt	Jewelry, Sculpture	Full Time/Day Time
Michael	Woodworking	Part Time/Night Time
Molly	Electronics, Microwave	Part Time/Day Time
Natalie	*Director of Education*	Teaching/Part Time/Night
Puppy	Leather Working	Full Time/Day//Night Time
Rob	Welding, Visual/Sculptural Art	Full Time/Day Time
Skunk	*Operations Manager*, Robotics Sculpture, Chopping Bicycles	Full Time/Night Time

there, as friends gathered more tools, they moved to bigger spaces, recruited more members, and the new members' interests in what tools and shops to have shaped the space to be increasingly diverse in regards to skills, expertise, and practices. Membership grew based on the founders' initial networks. These networks included a high number of people who regularly attended the Burning Man Event in the Black Rock Desert. Many of the values of Burning Man resonated at Artisan's Asylum: self- and community-reliance, cooperation, artistic expression, and freedom from mainstream social norms about pleasure and identity. At one time, the membership was so intertwined with the Burning Man event that the festival was the only time of year Artisan's shut its doors.

Spaces for Learning

In Figure 3.2, the network analysis map, Chris T., Brandon, Puppy, and Gabrielle are some of the biggest circles. Many people seek them out for help. Their studios are also in the center of the Asylum. Jess, the director of membership at the time and also organizer of rental spaces, told us that she thought strategically about where to place people in spaces, and when new ones opened up where people might move. She wanted people who were present often and who were active members of the community to be in the middle of space. She wanted people to learn from their examples as active members, and she wanted visitors to see that there were people using the space. In our observations, many of the studios on the fringes, in one of the other two buildings (also known as "the suburbs") were not occupied as often. Space, at the Asylum, was consciously constructed to encourage conversation, collaboration, and community.

How the space was arranged in other ways also encouraged interaction. Aside from a large common space with couches where people gather for lunch, invited talks, and just hanging out, there are low, four-foot walls around each of the studio spaces on three sides so that people can stop by and talk with one another, see what others are doing, and advertise what they, themselves, are doing. There have been occasions when people build up their space for storage and it gets to be a walled-in space that conceals what they are doing and they are asked to take their walls down because it is contradictory to the open and free-flowing communication that so many members value. In an interview, Ecco commented,

> I have access to a . . . essentially a wealth of continuing education in non-standard processes. You can throw anybody into a nice shop and be like, there's Joe, he'll talk to you when he has time but this is an education-based community-oriented environment where you're . . . we make the walls low so you can go up and be like, hey, are you the person who knows how to 3D print? Can you help me? Right . . . so like the whole . . . the inertia towards learning is lower than making almost in this particular environment. So, I have industrial processes which by themselves are very difficult to acquire out of pocket, welders and the ventilation required, the support tools, the space and then you have an alarming breath of experience present in the building who are essentially here on their own dollar. So you're invited and expected to interrupt, bother and collaborate with them. This is kind of a miraculous . . . this is like an oasis of thinking.

Artisan's affords its members the space, tools, materials, and in-house social networks to practice their crafts, to deepen them, and to learn new ones. Making spaces, Artisan's included, are often designed primarily as learning spaces. Learning happened in a number of ways: through the more formal classes, through one-on-one apprenticeships, but mostly through members talking with one another, bumping into each other, recognizing each other's expertise, and exchanging ideas. It is the layout, the hybrid communal and semi-private arrangement of workspaces that contributes to creativity and manufacturing at the Asylum. As a member of Artisan's, Chris, said, "The thing about this place, if you take a walk down the hall usually the idea you need comes to you, or the tool, or the material. But you have to be here, not just walking around outside" (Fieldnotes, 04/27). Learning to make at Artisan's means *being* in the space, not simply walking around, but spending time, interacting, and engaging intellectually and socially with others. It means contributing to the production of the space. The conditions of the Asylum, its qualities of networking, orientation to learning, pockets of different expertise and the possibility of some risk and failure (but without losing everything), make possible the learning-first orientation.

Interdisciplinarity and Expertise

Diversity of expertise was an integral part of communication at Artisan's. The Asylum was not just individuals sharing tools and a space. It was people who worked in different shops, used different tools, made different objects, and talked to each other. There was a large, interconnected, and dynamic network of people sharing information about expressive and functional fabrication. People shared ideas and collaborated with one another. That is, jewelers did not just talk to jewelers and mechanical engineers did not huddle together in a clique. There was a great deal of interaction between folks who spent time in different shops and had different forms of expertise. This heterogeneous network of expertise led to recognition of the specialization and expertise that different members had; they knew a lot about each other's work and strengths.

At times, people in different shops worked together because they had complementary skills. For instance, a long-time jewelry maker and the builder of the jewelry studio, Karen, worked with digital fabricators to laser cut acrylic molds for shaping softened copper into three dimensions. Mold making used to take two hours and it required careful skill, now the molds were cut in 30 seconds by the machine. Access to different tools and people with different expertise across shops allowed her to focus on her artistic vision.

Expertise, too, was relative. Art, an expert electronics hobbyist and maker, who made the first electronic skateboard for a hand-powered wheelchair that we ever saw, would go to Skunk for TIG welding advice. Skunk was the leader of the bike gang, SCUL, and the welder/creator of robots from outer space that he made from leftover bicycle and other metal parts. As one of the more skilled TIG welders, Skunk went to only one person for advice, a TIG welding teacher. The relative expertise of members in different fields is part of the reason we think of the network as dynamic and as people traveling along the lines of the network, not just at the nodes. Information and knowledge was not part of any one person, but was distributed, flowing from person to person through sharing. Those patterns of communication also led to hybridity in the kinds of work that were produced, and eventually, to the kinds of expertise and identities that people took on as individuals. For instance, Ecco began her artistic career as an illustrator, but soon branched out to 3D sculpture, kinetic sculptures, wearable art infused with electronics, and more. In part, that was possible because she was able to draw on the knowledge of the people who used the many different shops at the Asylum. The Executive Director at the time, Derek, explained the benefits of being at the Asylum:

> I think where we excel universally are in the overlapping edges of things. If you can find another place teaching jewelry that will occasionally take the jewelry class over how to use the laser cutter to make stamps for the hydraulic press, I'll be surprised. And I think if you look here, there is an opportunity for people to find like-minded, similar collaborators. But the exciting things are the, "Hey, I want to make prop lightsabers. Who can help me debug this LED driver?" and you get an electronics person to work with a prop maker. Or, "Hey, there's this opportunity to create these sensors for this thing," okay well we need a software [person]. We need an enclosure, manufacturing [person], a contracts [person], an electronics [person]. And those resources are all here.
>
> So I think the 3-Doodler, those guys absolutely knew what they were doing. They had years and years of making toys. They had everything lined up. They're like, "this is cool. I think people will like it. Let's have a kickstarter for marketing to get the tooling monies." And boom, they get a bajillion dollars. And the prototypes, they found somebody here to debug in the lab. And they could have outsourced that work. What is the value of them paying someone to sit in E[lectronics] & R[obotics] and debug a 3-Doodler motor noise or overheating issue versus finding someone equivalently talented, probably cheaper in Shenzhen [a city in China known for electronics manufacturing] or Nebraska? It's proximity. It's trust. It's interaction. [The 3Doodler inventor] could watch and learn and he can incorporate those into the next design he makes. So, I think if you're excited to see what your teammates come up with and enrich yourself, those are the kinds of teams here, because the teams here are teaching each other and learning.
>
> When we get an illustrator and a metal fabricator working together to make a sculpture, it's everybody bringing different parts of the puzzle and how those things result in new and exciting things. So I think those are the areas where you can't expect to find all the necessary expertise in a person. The resource pool here is very rich. But it still allows for direct collaborative cross-pollination in a team on a casual basis.

Ecco's 3D printer pen (the 3Doodler, a pen-shaped tool that extrudes heated plastic in a fine bead that cools to create stable structures) had been invented by a member of the Asylum. She had been one of the first people to test-drive it and received discounted or free pens when she needed new ones in exchange for the free advertising she provided through her exquisite works. At the same time, even individuals experienced in making things, toys in this instance, learned from the experiences of others, electrical engineers, who shared the same workspace. As Derek alluded to, many of those cross-pollinations were casual, or a product of serendipity. But that serendipity was crafted by the space, its tools, and those drawn to them in the same place and time. Other times, people actively sought out expertise from the community, and soon realized who might be able and willing to help them.

Collaboration, Communication, and Capital

Many people collaborated with one another. For example, while we were observing, Chris was asked to assist in making the casing of a data-probe for which two other members, William and Mike, had been contracted. William made the software and Mike designed and fabricated the electronics, but they needed someone to design and manufacture the casing. Given Chris's expertise, he was able to do that. Collaboration between these three independent contractors, essentially, was possible because of the diverse shops and expertise of the members. However, one of the bigger points Derek made was that people could learn from each other. The person who wanted to debug the motor noise was able to watch how another, more experienced, person did it, ask questions, and build up skills in a different area.

The member-run nature of the asylum means that members also have to depend on each other to accomplish tasks. Emily, the cartographer of imaginary places, had to ask a metal sculptor to help her construct a structure in her space. Someone kept unplugging her electricity, so she had to ask different people how she could get it back. Emily needed to talk to different people in the space just to be able to begin her work.

Artisan's maintains a number of email listservs that also encourage collaboration. One is "have-it-made" in which members, or those outside the Asylum, request someone to make something for them. Another is only open to members to discuss internal issues, such as governance, policy, grievances, or just to ask for help. Karen, an accomplished jewelry maker and teacher, was creating an art piece out of a metal newspaper box and needed someone to help her strip the paint from old one, something she had never done. By emailing the listserv, she was able to snag two volunteers. They not only volunteered their thoughts or opinions on how to strip it, they also volunteered their time and skills, skills that Karen needed to learn.

In an interview, Chris T. described a time he needed to accomplish a task: he was designing a buzzer for a neighbor of his who was to appear on the television show "Jeopardy Tournament of Champions" and wanted to practice hitting the buzzer. Chris had to design both a prototype of his project and a final design, once his client approved. He spoke to people face-to-face and emailed the listserv:

> I think the first thing I did was I reached out to Brandon here and I was like, hey, I'm looking . . . to emulate a keyboard . . . I think the question I posed was, I want to hit the space button on the computer electronically and we talked about various ideas and electronics we could build and how we could do it.
>
> And we talked about there has to be something out there that emulates a keyboard and . . . I posed that question to the mailing list here and [I] started getting feedback . . . and basically they already exist, a small circuit board that's known um, that was sold . . . called the Makey Makey . . . I have a prototype that's robust over here but it's gangly so I emailed the list, "Anybody has a Makey, Makey I can borrow?" I told them what the project was and S. gave it to me. Once I had the prototype and the Makey Makey, kind of that search for electronics was you know sitting around here asking a few people [who said] "Oh you could do it this way and there's three or four ways you could do it with specialized chips or you know like general purpose chips and I could make my own boards, spin my own board, like design it and everything," but it turns out they already exist on the market for seven dollars that's small enough on a circuit board and really the quantities we're talking about would be very hard for me to produce something like that.

Different people know of different ways to accomplish tasks and have their own resources. Together, these different outlets can be more useful than seeking one source of information. Gathering multiple answers to his questions allowed him to weigh his options and choose the one that suited his project best at different stages, the Makey Makey for the prototype and then dedicated processor chips for the final design. Chris found his information by being in the space and "sitting around here and asking a few people." While at the Asylum, one quickly realizes that expertise may exist in unlikely places, and so it is best not to assume who might know what. Bill said:

> When I first got here I spent about six months just building a robot, and every now and then I ran into problems, and I walked out and I would talk to Rob or somebody else. And just ask a question and just randomly, it was a very short length of time before I knew who to talk to about what. And I was getting high quality answers from people that I would never would have guessed.

After initially being open to talking to others, like Bill, people also figured out who to trust. Matt, a wood furniture maker said:

> I've been here for long enough now that I've kind of gotten to know, at least by face, if not by name, pretty much everyone who works in the

> wood shop with any sort of degree of skill. And I'll chat with them if I need something. If I have a piece of wood that I'm concerned might not serve the purpose I want to use it for I'll be like, "Oh what do you think? Have you ever done something like this before?" Sometimes it's no, I don't know. And sometimes . . . I think you can kind of tell when you're talking to someone in the shop if they know what they're talking about. And I don't know if someone's like, "Oh, interesting. That's a cool looking piece of wood. Maybe it'll hold together." That's different than if somebody's like, "Yeah, I don't know. Where does that crack go? Does that crack extend any further and is that along the grain? It's not too close to the edge, but I don't know." Those are the same ultimate answers, but they're very different.

Having something to offer others, whether it be helping to strip paint, knowing the capabilities of different electronics boards, or how a particular piece of wood will behave while sat on, is an important part of one's identity in making spaces. It is social capital that one can use to be recognized as a legitimate member of the space (Bourdieu, 1991). In some ways, this can be a drawback if you are a person who people constantly seek for help. However, it is also a positive because it means that when you have an issue that needs resolving, your own expertise will be recognized.

Social capital can be turned into access to information, but it can also be turned into actual capital. People earn money in many ways at the Asylum. Being at Artisan's is where many members are able to network and connect with others, that connectedness with the others in the community creates opportunities for them to make money. A few of our participants ran full-time businesses out of the Asylum. These businesses included Emily, an artist who made imaginary maps; Brandon, who produced secure microprocessors for running a local network of an internet of things; Chris T., who specialized in 3D printing and had a contract with the US Navy, and Ecco, a costume designer, welder, and artist specializing in hand-drawn 3D printed sculpture among other products (www.allthingsecco.com). At least one, a welder and artist named Puppy, based most of his income on work he did at the Asylum for others, including metalwork and teaching. Capital is built through competency and consistency. It also requires that one be comfortable enough with the other members to share oneself and one's work.

Becoming a Member, Learning to Communicate

Communication, here and in other places, is how people signal their identities. To what extent people claim to be part of groups or align themselves ideologically through the ways in which they present themselves to others, they signal the groups they are in and the groups in which they are not (Gee, 1996). But newcomers may not know which groups are "in" and which are "out" at the Asylum. There is not an extensive enculturation process as to how the Asylum operates as an interruption-happy oasis of collaboration. Derek spoke about what joining the Asylum is like for newcomers:

> Well, it's really hard to engage here. It's your first day of school, but it's not anyone else's first day of school. You're the new kid. And that can be tricky because people here are busy and powerful and talented and smart. And it can be off-putting.

Camila, recently moved from Brazil, was an architect/industrial designer who was creating a clean, sleek wooden keyholder. It took some time for her to figure out the kinds of acceptable interrupting:

> So in the beginning I wasn't talking that much because I was—I don't know. I was trying not to bother people. But then I started to know more people and now I see that I always ask things if I don't know—more like technical things about the woodshop and the machines. And also about how to make the packaging and stuff like that because I have this feeling that here people like to do things.

Derek was privileged to be able to spend time helping out and hanging out at the same time:

> For me, sitting at the front desk a few times a week for a year being able to ask so many people here, "Hey, how's that project going? Hey, what are you working on?" was just this huge benefit.

The way to become a part of the network of communication was to spend time in the space. It took some time for Camila to recognize that she was "invited and expected to interrupt, bother and collaborate with them" as Ecco had said. But, it also took what Derek was doing, which was contributing to the functioning of the space and spending time at the front desk. Spending time together, and not directly involved in their own individual work, was the way members got to know each other. These relationships built not necessarily around work but by "being in the space," as Chris K. pointed out, helped to form members' networks of expertise. The benefits of talking to each other were simple. Camila said, "I think when I started talking to people the projects improved."

Space, for makers, does not stop at the walls. Many present themselves and their products electronically as artisans or engineers. They use social media, create their own websites, vie for crowdsourced funding for projects, and sell their work on sites like etsy.com (we discuss online communication and presentation of self in Chapter 6). For members, although much of what they do is made possible by the space, the Asylum is not part of their outward facing identity as it would be if one worked for a larger company. It happens to be where they conduct their business. However, communicating with others in the space involves a great deal of identity presentation, willingness to learn and be humble, and contributions to the work of others, even if it means taking out the trash.

Relationships Between Space and Communication in Making Spaces

In this chapter we focused on the ways the production of physical, cultural, and social spaces are constituted through communication and the ways in which communication is mediated by space. We argued that there was an extensive network of expertise that people tapped to solve problems and that social capital was generated when the network recognized a person's expertise as contributing to the production of the social space of Artisan's Asylum (Lin, 1999). We also described the interrupting culture of the Artisan's Asylum, or the cultural values around interrupting other members while they are working to ask questions or seek guidance, and the emergent social and professional networks that result.

We have used networks to describe relationships between people and information at Artisan's Asylum. The use of networks is very much from the perspective of sociologists who try to understand the structures of organizations. We do not mean to reinforce the idea that structures govern what people do at the expense of individual agency. We are attempting to understand how people work in making spaces and that requires an understanding of how the organization works, its rules and divisions of labor, but also how individuals create the social space in which to enact making and literacy practices. At the Asylum in particular, individuals shaped the space.

At places like LMNPO, the social, cultural, and physical space was shaped by the leaders of the company. Communication was often, although not always, hierarchical. The leadership and design teams decided what everyone would do and the distribution of expertise was much more stark. However, at the Asylum, there was minimal hierarchy to shape the flow of communication. Flows worked much more through the multiple trajectories of expertise that the different shops encouraged.

Many high schools are thousands of students. Even the Asylum can only accommodate so many people at one time. The free flowing, volunteer, and honor systems may not be suited to compulsory education spaces. Still, the Asylum can offer some principles for thinking about how to create new making spaces and how to support communication.

To build spaces that encourage communication, making spaces can begin to create collaborative environments by posing facilities problems and communal responsibilities to members or participants. Solving those problems of maintenance together can encourage a culture of collaboration and foster joint decision making, leading to greater investment in the space.

Making spaces can also create space and time for learners to build relationships to others. Including social events, and structuring opportunities for young people to meet can help them develop the social trajectories of expertise that making networks can provide. In more formal settings, it requires rethinking how learners interact with each other. Instead of knowledge and knowing being the purview of individuals, as it is when educators assess learners through testing, knowledge becomes the domain of the group. Each person, although not responsible for making sure others know, is responsible for helping when they can and should also expect help when they need it. Finally, it reflects the work of those in STEM fields. Recently the National Science Foundation has called for scientists to work together on big ideas (2018). They encourage collaboration and cross-talk because, "As experts from different disciplines pursue common research challenges, their knowledge, theories, methods, data, research communities and languages become increasingly intermingled or integrated. New frameworks, paradigms or disciplines can form from sustained interactions across multiple communities" (NSF, 2018). To move STEM fields forward, those involved need to know how to communicate and how to contribute to the conditions that support that communication. They can learn from making spaces that encourage diverse simultaneous activity and give participants the opportunity and the structures to talk.

References

Bourdieu, P. (1991). *Language and symbolic power.* Cambridge, MA: Harvard University Press.

Buechley, L. (2013). *Thinking about making.* In Keynote presented at the Annual FabLearn Conference, Palo Alto, CA.

Chachra, D. (2015, January 23). Why I am not a maker. *The Atlantic.* Retrieved from www.theatlantic.com/technology/archive/2015/01/why-i-am-not-a-maker/384767/

Dougherty, D. (2013). The maker mindset. In M. Honey & D. E. Kanter (Eds.), *Design, make, play: Growing the next generation of STEM innovators* (pp. 7–11). Abingdon: Routledge.

Gee, J. P. (1996). *Social linguistics and literacies: Ideology in discourses* (2nd ed.). London: Taylor & Francis.

Ingold, T. (2011). *Being alive: Essays on movement, knowledge and description.* Abingdon: Routledge.

Lefebvre, H., & Nicholson-Smith, D. (1991). *The production of space* (Vol. 142). Oxford: Blackwell.

Lin, N. (1999). Social networks and status attainment. *Annual Review of Sociology, 25*(1), 467–487.

Massey, D. (2005). *For space.* Los Angeles, CA: Sage Publishing.

National Science Foundation. (2018). *Dear colleague letter: Growing convergence research.* Retrieved March 23, 2018 from www.nsf.gov/pubs/2018/nsf18058/nsf18058.jsp?WT.mc_id=USNSF_25&WT.mc_ev=click

Tucker-Raymond, E., Gravel, B., & Pierce, E. (in press). 80/20: Making identities in making spaces. In J. A. Gammel, S. Motulsky, & A. Rutstein-Riley (Eds.), *I am what I become: Constructing an identity as a lifelong learner.* Charlotte, NC: Information Age Publishing.

Vossoughi, S., Hooper, P. K., & Escudé, M. (2016). Making through the lens of culture and power: Toward transformative visions for educational equity. *Harvard Educational Review, 86*(2), 206–232.

4

IDENTIFYING, ORGANIZING, AND INTEGRATING INFORMATION

SETC Case Study

> **Focus Literacy in This Chapter**
>
> **Identifying, organizing, and integrating information** *Searching for information on the Web, using reference guides, material specification sheets, and product manuals.*

Introduction

In this chapter, we focus on the social practices youth enact to support their making at the South End Technology Center (SETC). SETC is a community technology center with a mission of engaging youth of color and members of the surrounding community in technology-rich STEM experiences that are authentic to their community and their lived experiences. A centerpiece of SETC's work is *Learn to Teach, Teach to Learn* (*L2TT2L*), a summer program that employs youth from around the city to learn STEM through technology-focused making, while also learning to teach younger members of the community about design, digital fabrication, and STEM-focused projects.

Within any making process, the search for, evaluation, and use of information relative to the goals of the project is centrally important to the maker's work. Identifying, organizing, and integrating (IOI) information is how we understand this treatment of information in the context of making. Information is everywhere, available in a wide variety of forms, from written texts to gestures, from photographs to material data sheets, from an artisan's skillful practices to procedural protocols. Practices for searching for, understanding, and using information are well documented in the literature, often conceptualized as information literacy (e.g., Mai, 2016), or digital literacy (e.g., Bawden, 2001). We seek to build on and to complicate these perspectives by situating IOI within making and STEM, to interrogate the ways individuals find and use the information they need for the particular purposes of making.

The multimodality, transdisciplinarity, and social characteristics of making spaces center information flow as critical to activities in the space. This flow of information is vital to the very existence and maintenance of making spaces. Modes for accessing and receiving relevant information are subtle, interpersonal, even idiosyncratic to the particular space and the members of its community. They are also specific to the tools, materials, and goals of the moment. By examining this flow of information through the lens of literacies, we provide descriptions of the multifaceted facility with representations that are tightly coupled with the activities and social relations of individuals within the space. We can then focus on how representations mediate this flow of information, and thus connect the literacies of making to the activities and goals of STEM. This chapter begins to describe the practice of IOI in a way that expands notions of information "literacy" to include the sophisticated social practices that youth, within a community of their peers, enact when making things that have meaning to them.

Identifying, Organizing, and Integrating Information

How one identifies, organizes, and then integrates the information needed to carry out design and fabrication activities is one of the core STEM literacies of making. Making is a multidisciplinary, multimaterial, and multimodal activity. Designing across platforms, approaches, disciplinary boundaries, and material familiarities requires facility with navigating and traversing entire worlds of information. To be good at making, one must be good with information. The maker must navigate the information needed to answer questions like these: What material would work best here? Can these two parts connect with each other? How have others solved this problem before me?

IOI is a particularly important literacy practice because the multiliteracies involved, including conducting research on- and off-line, are often the most difficult for young people to carry out with skill (Quintana, Zhang, & Krajcik, 2005). Finding and managing information is central for engineering design (Fosmire & Radcliffe, 2013), and we posit that skilled makers know how to find information from others, evaluate it, recontextualize it for their own research questions, and figure out what else they need to know. Where to find parts and materials, how to connect electronics to microprocessors, and constructing a particular part using a new technique are all activities that require finding and using information. We hypothesize that this is a productive place to begin finding points of connection between experienced makers' practices, youth practices, and the kind of work required of making in and out of school.

In a prior work (Gravel, Tucker-Raymond, Kohberher, & Browne, 2018), we reported findings from interviews with experienced makers on how they engaged IOI in the different processes of making. We found, first, that makers use information from a wide array of sources, including web pages, books and articles, images, patent documents, manuals, and conversations with peers, both in person and online. In these interviews, makers described their processes of ideating, designing, and producing some artifact in ways that suggested IOI operated with specificity in each making process. Makers used online image searches for inspiration, and they examined spec sheets to determine the compatibility of different parts. They found information on the Web and from friends to help reconcile issues. Across all of the accounts from experienced makers, information was the life-blood of making, existing in many different forms. This flow of information is a central component of the networks of communication and knowledge sharing that exist in spaces like Artisan's Asylum and the South End Technology Center.

To understand the flow of information in a youth community making program, we explore socializing as a resource for making (Ito et al., 2009). Youth socializing is a ubiquitous feature of "hanging out," yet it is often considered "off-task" or "taboo" behavior in formal learning settings. We build from Larson and Gatto's (2004) work describing the "tactical underlife" to understand youth "tactics" (de Certeau, 1984) for operating within STEM spaces. de Certeau (1984) argued that—within disciplined spaces, like STEM, where power is produced and reproduced through particular conventions and norms, like representations—"ordinary people evade the sociotechnical mechanisms of regimes in their efforts to 'make do': to get the best 'wins' they can" (Knobel & Knobel, 2002). Individuals create space within the sometimes inhospitable institutional environments—i.e., dominant practices and conventional technical representations of STEM disciplines—to make sense of their activity and to participate in their own ways. de Certeau (1984) refers to these moves as "tactics" and positions them as opposing the "strategies," or the institutional and ideological practices, that may be valued within STEM disciplines (e.g., reading a circuit diagram to understand how an electronic component works). Tactics are "the ways the people who live within institutionalized spaces use, manipulate, and divert the space to which they are assigned" (Larson & Gatto, 2004, p. 14).

In this chapter, we focus on a youth summer program in a community making space, South End Technology Center. We describe how the program itself was designed to challenge dominant and exclusive ideologies around STEM learning, where new makers worked alongside experienced makers in a cascading mentorship structure that supported their making. The ways youth participated and developed relationships with each other became part of how they identified, organized, and integrated information in their work. We present two episodes of youth making at a round table that was a hub of social activity in SETC. By focusing the chapter on that particular location within the space, we can begin to uncover the nuance and sophistication of the ways in which youth practices, including the role of socializing and hanging out, supported their STEM-rich making.

Our Approach to Studying SETC

In our study, we were able to conduct intensive observations at SETC, capturing more than 200 hours of observations in fieldnotes, video recordings, and audio recordings. One methodology we employed was the use of a head-mounted video camera. We asked youth sitting at a round table at SETC, where another seven to eight youth generally gathered, to wear the camera as they worked. This produced over three hours of recorded moments. We describe the SETC philosophy, introduce the participants, and then share analysis of two instances from the round table illustrative of the ways socializing and navigating information interacted as the youth worked on their projects.

The South End Technology Center

The South End Technology Center was founded by long-time community activist and MIT faculty member Mel King, Hon.D. It houses many programs to support the community's engagement with technology in Boston's South End. SETC seeks to support people of color in experiencing STEM and technology-rich activities in powerful ways that relate to their communities and identities. One of the flagship programs is *Learn to Teach, Teach to Learn (L2TT2L)*. For more than 15 years, this youth summer program has prepared young people of color from around Boston to learn through making with technology while also serving as teachers for younger members of the community, exposing them to technology, design, and engineering activities at SETC and around the city.

SETC pursues empowerment for communities of color. The work of SETC, and *L2TT2L*, is organized around a core principle: *technologies of the heart/technologies of the earth*. In an interview with King, he described this foundation:

> I've always focused on two words (heart/earth) and I would start with the bigger goal, the story about turning swords into ploughshares and being at MIT and [I was] challenging some of the ways in which they talked about high tech I used to say, "Anything that takes life is low tech. High tech enhances life." And so you'll hear all these folks from engineering [and] science [talking about high tech weapons], and if you listen to what [the phrase from the bible] said, you'd beat your swords into ploughshares and your spears into pruning hooks.

Technologies of the heart focuses on how people can come together, to build relationships, and through those relationships, solve problems in the world and in one's community through the design and use of technologies. The work is rooted in historical freedom struggles for African Americans and other people of color.

Youth Teachers work in groups to identify problems in their communities and design solutions. They are supported through a cascading mentorship structure, where each group has a College Mentor, usually someone who participated as a Youth Teacher in the past. There are also adult mentors from area universities and the community. Youth Teachers are also expected to mentor younger children by teaching at camps and community centers throughout the city. Participation is structured in such a way that almost everyone is both a mentor and a mentee. For those whose social languages are often excluded from learning spaces, the cascading structure helps to provide access to learning STEM content. The cascading mentorships are meant to help Youth Teachers learn from more knowledgeable people. However, the near-peer character of the cascade means that those in different roles (college mentor, youth teacher) also speak similar social languages. The cascade thus functions as a scaffold for supporting multiliteracies, for moving between social and technical languages, and at times creating a hybridity of multiliterate social practice that functions to further learning.

The Participants

The year we conducted our research at SETC, there were 9 "College Mentorz" and 36 "Youth Teachers": 50% of participants identified as girls and 50% as boys. About 55% percent of youth identified as Black or African American (this included young people who were 1.5 generation immigrants from Haiti, Nigeria, Ethiopia, and Ghana); 30% identified as Hispanic or Latinx; 9% identified as Asian or Southeast Asian; 4% as Arab, and 2% were white. The Youth Teachers were all high school students, most of whom had little to no prior experience with digital fabrication, physical computing, or making with technology.

The Space

SETC occupies two units in a mixed-income housing building in Boston's South End. The space consists of three main rooms: a computer lab, where founder Mel King can be found almost every day at his desk, a central workshop room with computers around the periphery, and the Fab Lab. In addition to the Fab Lab and workshop described in the introductory chapter, there is an alcove where a large set of shelves and a piano sit just on the edge of a large round table. The table sits eight students comfortably, and every day that we were at SETC, youth were gathered around this table working on their projects. Music was played from computers at this table. Youth talked, joked, asked each other questions, and enjoyed each other's company. Others would stop by to talk with those already sitting. Two College Mentorz sat there nearly every day. The table was a hub of activity, socialization, and a critical site for information flow at SETC.

IOI at SETC

Youth Teachers were introduced to different sources of information. During the spring months, Youth Teachers and College Mentorz met regularly on Saturdays to learn specific technologies, like physical computing with Arduinos, Scratch coding, and digital design and fabrication with the vinyl cutter. Within each of these sessions, Youth Teachers were introduced to numerous sources of information. These included websites to support coding, like the Scratch Community or Codebender.org, online electronics distributors, like Sparkfun and Adafruit, and useful channels on YouTube. College Mentorz, having gone through the *L2TT2L* program as Youth Teachers in the past, were knowledgeable about specific sources of information related to particular tasks, like where to look for LEDs or how to find help writing code for the Particle Photon. Like we saw with experienced makers in the analysis of interviews, information was identified across a wide variety of sources and forms (e.g., written texts, images, videos), but information was also found in talking with others in the making space. Peers and mentors at all levels were sources of information.

What follows are two episodes of youth making at the round table. In each, we explore how socializing and the search for and evaluation of information were intertwined. We present the episodes to support the argument that youth practices for finding the information they need in making are complex, multimodal, and inherently social. This has implications for how spaces are structured to support IOI as a social practice, where the community and its network enable the flow of information in making.

Episode 1: Sourcing Parts and Information: Luis Finds an LED

Luis sat at the round table next to eight other youth makers. Music was playing from a laptop. Luis asked Somtoo if she had work later. She complained that her legs were itchy. He asked if she had bed bugs, and others laughed. As they were talking, and hanging out, Luis searched for an electronic component for his group's design: a digital device to help one locate a lost cell phone, keys, or wallet. He conducted Google searches, using keywords like "RGB" and "LED." He located a LED that was likely to work for their project on the Adafruit website (see Figure 4.1). He scanned the images on the webpage, clicking on each of the pictures multiple times (Figures 4.2–4.5). He appeared to scan past a technical diagram (Figure 4.6), containing information about

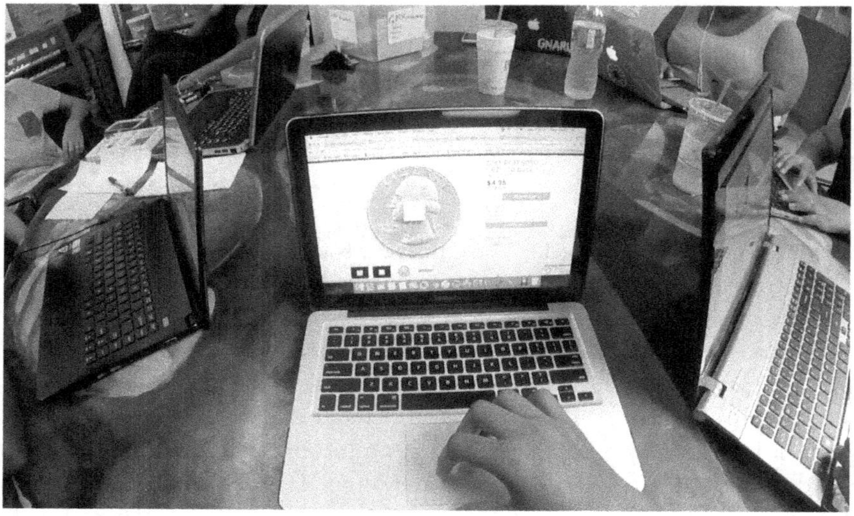

FIGURE 4.1 View from Luis's head-mounted camera, where he examined a webpage for the LED he found for his group's project.

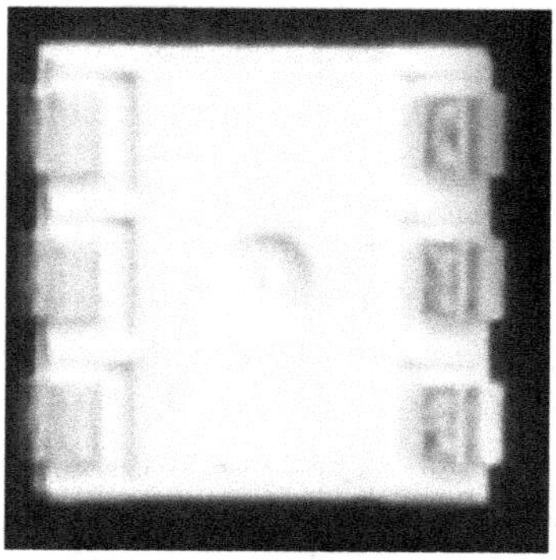

FIGURE 4.2–4.5 Images from the webpage for the surface-mount LED, showing the front, back, its relative size, and what the package of 10 LEDs looks like.

FIGURE 4.2–4.5 (Continued)

power needs and a wiring diagram. Turning to Keegan, Luis said, “Look how small that is. It’s really small. Pack of ten? But I don’t know what it is. [reads from the electronics website] ‘These surface-mount LEDs are an easy way to add a lot of colorful dots to your project,’ which is exactly what we need! How’s it work, though?”

Luis’s comment to Keegan is an invitation to explore what he found. It also served as an announcement that he did not know how the LED worked. Less than one minute later, Luis spoke up again, with another announcement to the group:

LUIS: I have no clue what I’m doing.
HAHN: Don’t say that!
SOMTOO: Remember that time I took my final and I called you [Luis] to help me?

TECHNICAL DETAILS

- Red Vf: 1.8 to 2.1V
- Green Vf: 3.0 to 3.2V
- Blue Vf: 3.0 to 3.2V
- Red color: 620-625 nm
- Green color: 520-525 nm
- Blue color: 465-470 nm
- Red brightness @ ~20mA: 600-800 mcd
- Blue brightness @ ~20mA: 800-1000 mcd
- Green brightness @ ~20mA: 1500-2000mcd

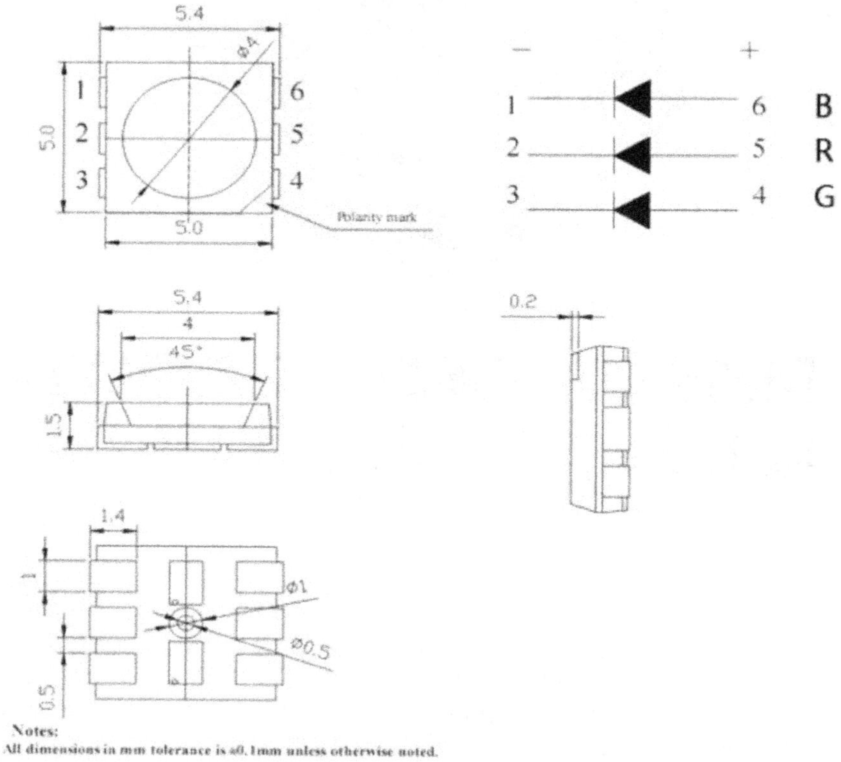

FIGURE 4.6 The "Technical Details" offered on the product website for the LED that Luis found.

Luis exhibited a kind of self-doubt, first by asking how the part worked and then by saying "I have no clue what I'm doing." Prior to saying this, he conducted multiple Google searches using terms like "RGD," "LED," and "hm10" (a protocol for Bluetooth). He arrived at a LED for sale on Adafruit's website that could be successfully integrated into their project. And yet, he positioned himself as having "no clue." Hahn and Somtoo challenged his declaration, and Somtoo offered evidence to the counter that one time he helped her in school, as a college student no less, and he did have a "clue."

We interpret Luis's comment as a "tactic" (de Certeau, 1984) for two reasons. First, his expression of self-doubt, positioned him as "not knowing," which is a move exhibited by many youth to distance themselves from situations where perhaps they are uncomfortable with the material, or feel confused about something they see. The technical diagram (Figure 4.6) included in the product information might be able to offer some information relevant to his quest. However, the "strategies" employed by more experienced makers (de Certeau, 1984) to read such specific and conventional notation is something Luis was still learning. Seeing that diagram, and the relatively technical presentation of information about the LED, may have cued his use of the tactic as means of subverting that inhospitable representation. Second, his comment also seemed to invite those around him into conversation about the information he was trying to understand. He said, "I have no clue" as if to entice Hahn, Somtoo, or others to engage with him in making sense of the part. And they did.

During the episode, as he worked, he constantly talked with others at the table. They discussed music, lunch options, and made jokes. Within that socializing Luis made space to draw in help in navigating the information in front of him. Following the previous exchange, Somtoo walked around the table to see what Luis discovered:

SOMTOO: Ohhh, what is that?

LUIS: I don't know. I think it's an RGB. But I don't know how it works.

SOMTOO: It turn red, green, and blue, right? What do you mean you don't know how it works?

LUIS: No, like . . .

SOMTOO: You plug in the things, the way they gotta go, to turn it on.

Somtoo demonstrated interest in what Luis was doing, she was a College Mentor, and part of her responsibility was to support the Youth Teachers. She appeared confused that Luis said he did not know how it worked. We understand his response, "No, like . . .," to mean he has something to say about RGB LEDs, but the specifics of this component, and whether it would fulfill the requirements of their design, were in question. The technical diagram displayed on the website contained the answers to Luis's questions, however, Luis keenly navigated this space using his socializing and ways of connecting with others around the table. Somtoo went on to answer the question we believe Luis was asking:

SOMTOO: You wouldn't use a battery pack with it. You would put like, you know, the 9 volt? You put a little clip on that with the two wires and plug 'em in. And that's how it would turn on.

TYLA: Yeah.
SOMTOO: 'Cause when we showed you guys this stuff, we had the little, the one that looked like a coin, and you just, like, slipped it on.

Tyla, another College Mentor, confirmed Somtoo's description, providing Luis with the information he needed to make a decision about the LED. Luis concluded the episode saying, "So, that works," and he entered the information about the LED in their group's parts ordering spreadsheet. By engaging Somtoo's help in evaluating the information in front of him, Luis determined that the LED component he originally identified would work for their design; and, in fact, it did. After adding the information to the spreadsheet, Luis turned to Luisa, to show her the part (using Figure 4.4). He said, "Look how small that is?" Luisa replied, "Holy sh*t!" He maintained the socializing they were doing all throughout the episode, sharing things with each other and just hanging out while still finding out information to move his project forward.

Episode 2: Making Sense of Mistakes: Mateo "Blows up" a Motor Controller

On this particular day, the camera was mounted on the wall overhead, directed at the same round table. Nine people were working with computers and electronics. Music played from one of the laptops; it was loud on this day. People were at their computers. They were writing, searching, or coding, and they talked as they worked, about food, about the songs, about things that happened outside of SETC. There was a lot of laughter and a lot of banter. The music kept getting turned louder. They were socializing, hanging out, and working.

Mateo was part of a team developing a remote control car gaming system that could teach youth about coding and physical programming; they said they wanted to build something that would introduce kids to coding, like what Scratch has done. Mateo stood, back to the camera, hunched over a breadboard with some electronic components, a three-cell battery pack, and an Arduino. He used jumper wire to connect the components. The music played as Mateo inserted one last wire; a loud "snap!" was heard, and he jumped backwards. Freddy and Somtoo broke into laughter. They were laughing hard. Freddy reenacted Mateo's reaction as he laughed, pretending to jump backwards. Somtoo told him to disconnect the power supply before doing anything else. Mateo did not respond to her, but simply said, "Damn, that motor blew up!" Freddy asked, "what's that smell?" Mateo replied, "It's smoke!" Somtoo added, "Mateo, you messy!"

Mateo seemed a bit shaken by this and a bit afraid of what might happen next, in part because he was not entirely sure why the "snap" happened:

MATEO: How many volts is this?
FREDDY: How many volts are you right now? [Freddy breaks into laughter]

MATEO: No, I didn't touch it!
SOMTOO: He connected it.
MATEO: Yeah.
SOMTOO: 'Cause he touched, he was sshhh (she pretends to fall backwards)
MATEO: Then I was just (re-enacts falling backward).

Mateo began by asking a question about the battery pack. Freddy turned the question into a joke, but Mateo clarified what happened, that the event was caused by connecting some wires. The group laughed with Mateo and he laughed too. He examined the parts and noticed that it was the motor driver that gave off the smell, not the motor itself. Someone else walked over, and Mateo shared that he "blew it up!" He remembered there were more motor drivers in the other room, and he turned for the electronics bench in the Fab Lab to get another one. As he walked away, Tyla, also sitting at the table, said, "Doesn't mean you can blow them all up!"

When he returned, Keegan was sitting at the table. Mateo joked with him, "Can you grab that?" pointing to the motor controller; Freddy laughed. From his reaction to this moment, and what followed, it seemed Mateo was trying to make sense of what happened. He appeared hesitant to touch anything, looking closely but not putting his hands on the components. Asking Keegan to "grab that" was a good-humored way of avoiding touching the components. Mateo was still unsure about what caused the "snap," as though he was wondering if there were still a lot of "volts" there, and perhaps it could happen again if he touched it. Mateo said, "It blew up real good!" To which Somtoo replied, "Didn't I tell you what to do the other day? Put some resistors on that. And everyone like 'Noooo, noooo'." The group engaged in a light hearted argument about whether resistors are helpful. Mateo said they didn't matter, and Tyla and Somtoo, the College Mentorz, argued they were important. Few specifics were offered, but the relative positions were made clear. The argument ended with Tyla and Mateo exchanging childish grunts toward each other while smiling, without clear resolution of the question about resistance.

People got back to work. Mateo removed the batteries from the holder. He held one of them between his fingers, in different configurations (see Figure 4.7).

After doing this for a minute or so, he raised a question:

MATEO: I have a question. If I go like this (Figure 4.8), why don't I have this circuit blow up?
SOMTOO: Because you're not adding no energy.
MATEO: This battery's adding energy . . . why won't I short circuit . . . if I'm a conductor (Using a finger from his free hand, he gestures a circle around the loop he has created with the battery and his fingers.)
SOMTOO: That's resistance!

MATEO: I understand if I go like this (Figure 4.9). But if I go like this (Figure 4.8) . . . that should just go like this [same gesture, around in a circle].

SOMTOO: You just did the same thing. Are you grounded?

MATEO: Is there a ground?

SOMTOO: Are you grounded? Cause I get what you're saying, but you putting just one hand here . . . and like.

MATEO: Oh, no, I'm just kidding. [pause]

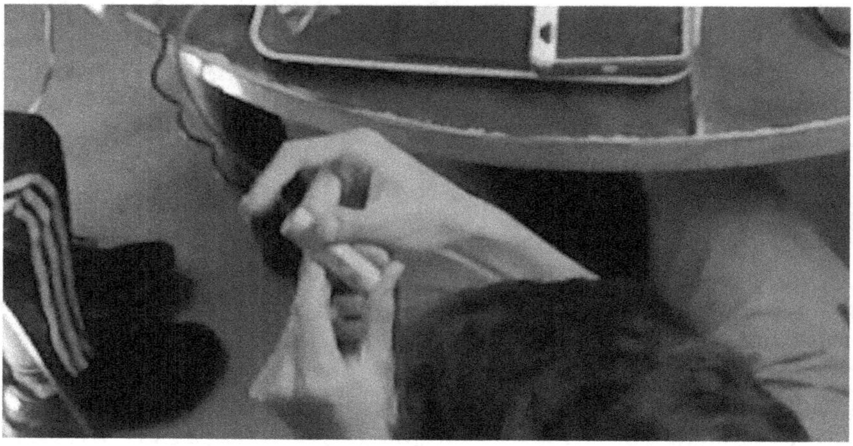

FIGURE 4.7 One of the configurations Mateo tried as he quietly explored different ways to holding the battery between his fingers.

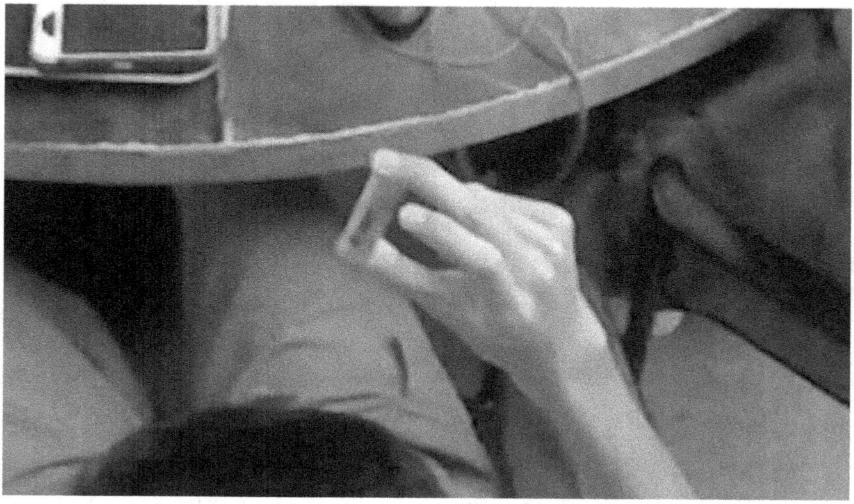

FIGURE 4.8 The configuration of the battery Mateo held when asking for help understanding why this would not "blow up," like the motor controller had.

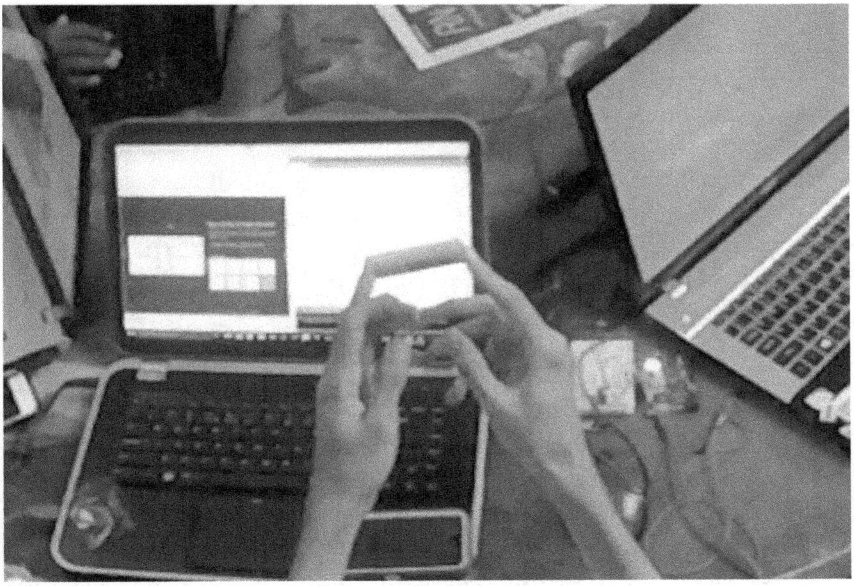

FIGURE 4.9 The example of holding the battery in a way Mateo does not believe should "blow up."

Mateo asked a question about what creates a short circuit, presumably because that is how he understood what happened to the motor controller. Clearly he still wondered what had happened. Somtoo and Tyla tried to argue that the circuit needed resistance, but Mateo's persistence on this issue suggested that their answer did not settle his curiosity. He had not yet found the information he needed to make sense of the incident that occurred just a few minutes earlier. Mateo demonstrated two different ways that he could hold a battery, both where it would not "blow up," like the motor controller had. Somtoo offered, "That's resistance," perhaps referring to the fact that his body acted as resistance, and thus the short circuit that led to the motor controller "blowing up" was not happening with his fingers because of the resistance in his skin. She then asked whether he was grounded, the reason for which is not clear, but Mateo went along, "Oh, no, I'm just kidding . . ." as though to imply he now understood what Somtoo was saying. Her question about a "ground" could have been interpreted as a rationale or an answer. He paused, letting that comment stand as though Somtoo's mention of a "ground" made sense to him. However, Somtoo looked at him continued:

SOMTOO: You didn't fool nobody! When you say that! You doing this, and this is the same thing (she makes the same gestures with her fingers as Figures 4.8 and 4.9). Cause it's your body. Like your hand is involved here.

MATEO: Well, no, that's like saying putting a wire like that is the same as putting a wire like this (he mimics having wires of two very different lengths with his fingers). That's a little bit more resistance. No, cause . . . it's like R . . . no, resistance equals rho times lambda over A.

SOMTOO: Cause you're not anything, it's not charging, it's not powering . . . why would you short. But you're not taking anything from there [the battery]. Like, you're conductive, but that don't mean it's going to give you anything.

MATEO: If you put two wires, they're not having any energy, but they're still going to . . .

SOMTOO: You know what? I'm not even Google, so I'm just going to help you out (Somtoo goes to Google on her phone to find resource).

At the start of this exchange, Somtoo seemed to realize that Mateo still had questions, and she continued her explanation. Mateo argued back, saying that holding the battery between fingers on one hand (Figure 4.8) was similar to holding the battery between fingers from two different hands (Figure 4.9). He compared that to a circuit with a short wire and another with a longer wire. Somtoo argued that nothing is being drawn from the battery, acknowledging that Mateo is conductive, but that does not mean electricity will flow. Mateo still did not seem satisfied, but Somtoo ended the conversation turning to Google on her phone, as if to suggest the Internet could provide resolution.

We interpret Mateo's questions as his search for meaning around the earlier comment that the circuit needed resistance; this meaning was important information in his pursuit as a maker. As Mateo placed the wires in the breadboard, connected to the power supply, we can imagine from his reaction that he did not assume the controller would "blow up." When Mateo was told that he needed resistance, he seemed to doubt that, and he continued to make sense of the situation.

Mateo engaged Somtoo for help understanding the problem. After she turned to Google, she shared that people had been hurt playing with really powerful batteries, suggesting that Mateo's demonstrations would produce different results if he used a 12v battery, like one used in a car. Her response, in a way, validated his curiosity about why one's body, which is conductive, would create the same kind of short circuiting that he experienced with the motor controller.

Through both examples, youth worked through issues they encountered by seeking information from each other and online. They socialized throughout their work. They joked with each other, laughed at and with each other, listened to music, and generally enjoyed each other's company as they worked. Much of this behavior could be characterized as off task or even inappropriate, depending on who observed it. Had an authority figure overhead Freddy laughing at Mateo or saying, "How many volts are *you*?" we can imagine some disapproval. And yet, Mateo operated within this moment to enlist the support of his peers in making sense of the incident that he just experienced. The information he needed to answer his questions, the reasons why it blew up, was available in various forms on the Internet or in books; there are posts about motor controller circuitry, and Adafruit.com has a module on skin conductivity with batteries. Information exists within the scope of sources that we observed makers using in their work. However, Mateo deployed a tactic in this moment. By playing into the ways that others made fun of him, and by pretending to understand ("Oh, no, just kidding"), he invited Somtoo as a sense making partner. She became a source of information that he identified and that worked to understand.

Information Flowing Through Meshworks

At SETC, the *L2TT2L* program provides a structure for youth of color to engage in STEM-rich making. This structure includes carefully designed supports like cascading mentorship and opportunities to relate making to social issues in the youths' lives and communities. Within this making space, we were able to observe how youth use and develop literacies to engage in STEM. And, in this chapter, we focus on the ways information flows through the community, and how that information is navigated as youth solve problems, technological or otherwise.

The two episodes in this case are both examples of youth deploying tactics to gather information and support while working to resolve questions, all within the context of socializing around the round table at SETC. Throughout the program, they were introduced to different resources (e.g., Adafruit, Arduino coding supports, electronics workshops) during their making and in response to particular issues or roadblocks that emerged. In these moments of challenge, when things did not go as planned, youth tried to make sense of what was in front of them, be it a piece of information or the behavior of an electronic component.

The episode with Luis illustrated the challenge of sourcing components in making, where conventional representations are a common way the distributor shares information about the technical specifications of the part. Luis enacted what we consider to be a tactic to find ways to work within the "inhospitable" environment of the "technical details" diagram (Figure 4.6). He scanned the images of the part and enlisted the help of those around him, those he was socializing with at the time, by claiming he did not know what he was doing. He was trying to evaluate the component he identified in a Google search, and, together with his peers, he determined that the part would work in their design. He placed that information in a spreadsheet so the part would be ordered and integrated into their design.

The episode with Mateo illustrated how incidents occur in making that youth want to understand. For Mateo, the motor controller "blew up," and he wondered why. Within the socializing that existed at the round table, he enlisted his peers to help him make sense of this event, even relating it to an analog situation with his fingers and a battery. While he may not have resolved the issue completely, it was a moment of Luis making sense of his work, engaging ideas in STEM with his peers, as an example of how his literacies helped him gather and evaluate information.

Youth's ways of practicing IOI involved complex relational structures, using socializing, humor, and indirection as modes of exchanging information and critique (Wright, 2016). While we can only describe the atmosphere and conversation at the round table in so much detail here, Somtoo, Mateo, Luis, Freddy, Tyla, Luisa, and Hahn knew each other well, both through their time at SETC, and from other interactions in their lives (e.g., Somtoo referencing Luis helping her and someone else in school). Their ways of being with each other around this table, talking life, laughing together, and socializing with each other, built their relationships, both around friendship and around interests (Ito et al., 2009). And the practices they enacted to construct these relationships supported the ways they navigated information needed in making.

When it came to identifying information, youth used a similar array of sources as compared with the experienced makers we observed: YouTube, texts, images. However, we understood IOI as more of an individual practice with those we interviewed at Artisan's and LMNPO (Gravel, Tucker-Raymond, Kohberger, & Browne, 2018). We found IOI was a crucial practice for experienced makers, where they constantly utilized a vast knowledge of sources and schemes for organizing and integrating the information they needed in their work. They possessed certain strategies for deciphering relevant information contained in specific kinds of representations, like the technical details diagram (Figure 4.6). Yet, information also flows differently at SETC, with youth makers. The relationships of the space and the dynamic network of individual interactions became the matrix for how information was both identified—e.g., Luis assessing a LED's functionality—and evaluated. The community worked to understand different sources of information that were then organized and integrated into different projects as needed. Ultimately, it was their ways of communicating with each other that allowed information to flow around the table, and this positions IOI as a social practice that the members of SETC enacted within the dynamic social networks in the making space.

Chapter 3 outlines how communication was mediated by the physical, social, and cultural spaces of Artisan's Asylum. Communication that built and maintained a complex and dynamic social network. This network was the infrastructure for the flow of knowledge and information. Both people and information flow through making spaces, making contact with others at certain times, missing each other at other times. This flow is dynamic and fluid, built on relationships cultivated within the structure of the space. Because the membership at Artisan's is multidisciplinary, comprised of individuals with a diverse range of interests and skills, communication allows information to flow across disciplinary boundaries in ways it might not in more disciplined spaces, such as a pottery studio or an engineering firm. At SETC, a very similar kind of dynamic social network is developed by the Youth Teachers, the College Mentorz, and the adults who construct that space through the cascading mentorship structure. At the same time, youth mostly work in groups, together in the Fab Lab or around the circular table, forging relationships through socializing and constructing and maintaining that network.

Socializing, or the ways youth connect with each other as friends or around interests, becomes an integral element of how youth practice IOI, as it contributes significantly to the constellation of relationships they establish and maintain. That network, and the ways they populate it, becomes the infrastructure through which they seek, evaluate, and use information in making. These means of traversing worlds of information socially, together with others in the space, are how identifying, organizing, and integrating information take form. Identifying comes through how youth notice and focus on problems, confusions, or curiosities: saying "I have no clue" to signal the need for help in assessing new information. Organization comes in many forms, through discussions to sort out these confusions, and through recording decisions in a spreadsheet (e.g., once Luis decided to order the part he found, Tyla told him to add it to their wish list spreadsheet). Perhaps most interesting is how nuanced, specific, and personal the integration of information appears to be. If we think of integration as both sense making and decision making about materials and processes then IOI becomes as much about sense making as it is about managing information.

If IOI is practiced within networks that are dynamic and fluid, educators and researchers need new tools to help understand these flows, and the ways the networks are sense-making infrastructures for individual agents. One possibility, as we raised in Chapter 3, is to reconceptualize networks as "meshworks" (Lefebvre, 1991). Meshworks focus not on the nodes—the people and what they know—but rather on the relations and the lines of life that individuals live as they move in and out of different spaces. As one travels these multiple lines of life, relations are cultivated over time as the lines intersect with others. These intersections, or knots (Ingold, 2011), combine different lines with those of others in particular moments to achieve some tasks, settle some dispute, resolve some tension, or to simply get to know each other. Thus, the meshwork is a web of intersecting lines within a space, produced and maintained by the multiple actors who inhabit it. As people come and go, as the space is constructed and reconstructed, the meshwork shifts and changes to reflect new lines, new knots, and new ways that individuals relate to each other through the work. Meshworking captures the dynamism of being in the world, and how one learns to seek and use the information one needs. It helps us understand the underlying infrastructure through which, and within which, individuals are able to get the information they need for the tasks they wish to accomplish, like designing and making technological solutions to problems in their communities.

One of the characteristic features of making is that there are always many ways to achieve a goal. And whichever way emerges as the chosen one is the result of the maker gathering the necessary kinds of information needed to make that determination. For youth at SETC, socializing is a major part of how the information flows, and thus how decisions are made. Furthermore, the cascading mentorship structure introduces a range of experiences and familiarity into the meshwork, where newcomers and oldtimers are in conversation. Their socializing enriches the texture of that meshwork, knotting lines together that connect a broad range of experiences. The structure at SETC encourages socializing, and it becomes a space where youth "tactics" are resources for finding the information they need. This means IOI operates within the meshwork, contributing to moments of sense making around STEM-related information. We argue that as youth encounter different representations, their tactics and the support this meshwork provides places them in position to confront and make sense of conventional and protected representations. The "technical details" diagram can become another source of information that collectively youth can interrogate and understand. Thus, youth participation in making spaces like SETC could provide avenues for IOI to transform how youth find, evaluate, and use new kinds of information represented in different forms, where their ways of "making do" and hanging out are considered resources for learning.

References

Bawden, D. (2001). Information and digital literacies: A review of concepts. *Journal of Documentation, 57*(2), 218–259.

Cole, M. (1996). *Cultural psychology: A once and future discipline.* Cambridge, MA: The Belknap Press of Harvard University Press.

Cope, B., & Kalantzis, M. (2000). *Multiliteracies.* London: Macmillan.

De Certeau, M. (1984). *The practice of everyday life.* Berkeley, CA: University of California Press.

Gravel, B. E., Tucker-Raymond, E., Kohberger, K., & Browne, K. (2018). Navigating worlds of information: STEM literacy practices of experienced makers. *International Journal of Technology and Design Education, 28*(4), 921–938.

Fosmire, M., & Radcliffe, D. F. (Eds.). (2013). *Integrating information into the engineering design process.* West Lafayette, IN: Purdue University Press.

Ingold, T. (2011). *Being alive: Essays on movement, knowledge and description.* London: Routledge.

Ito, M., Baumer, S., Bittanti, M., Cody, R., Stephenson, B. H., Horst, H. A., . . . & Perkel, D. (2009). *Hanging out, messing around, and geeking out: Kids living and learning with new media.* Cambridge, MA: MIT Press.

Knobel, M. & Lankshear, C. (2002, February). *Steps toward a pedagogy of tactics.* In Keynote address at the Annual Conference of the National Council of Teachers of English Assembly for Research, New York.

Larson, J., & Gatto, L. A. (2004). Tactical underlife: Understanding student's perceptions. *Journal of Early Childhood Literacy, 4*(1), 11–41.

Lefebvre, H., & (1991). *The production of space* (D., Nicholson-Smith, Trans). Oxford: Blackwell. (Original work published 1974).

Mai, J. E. (2016). *Looking for information: A survey of research on information seeking, needs, and behavior.* Bingley: Emerald Group Publishing.

Quintana, C., Zhang, M., & Krajcik, J. (2005). A framework for supporting metacognitive aspects of online inquiry through software-based scaffolding. *Educational Psychologist, 40*(4), 235–244.

Wright, C. G. (2016). Constructing a collaborative critique-learning environment for exploring science through improvisational performance. *Urban Education.* doi:10.1177/0042085916646626

5

FROM FLUENCY TO FLUIDITY

Representations in Making

Focus Literacy in This Chapter

Creating representational forms and traversing representational systems and materials *Producing symbolic representation of any aspect of the project as an aid to making (measurements, notes, drawings) or transitioning between different modalities such as talk, sketches, and formal designs.*

Introduction

Central to forms of work, activity, and learning in STEM spaces are the representations that individuals use to create, communicate, and reason about their worlds. Considerable research has been conducted on the many roles and forms of multimodal representations in mathematics, science, and engineering. Our focus in this chapter is on representational fluidity, or the ways in which people create and traverse representations in the various processes of making.

We argue that instead of representational fluency, or the ability to read, use, and create a particular kind of representation, we can better recognize and support learning through the concept of representational fluidity. Fluidity is intended to capture the dynamic nature of these productions and traversals, while also acknowledging and blurring the lines between the variety of idiosyncratic and conventional representational forms in making. We highlight the unique and relatively unstructured nature of moving across and through representations while engaged in making, from ideation, to design, to iteration. We define this literacy practice as using and producing symbolic representations in any aspect of the life of a project. Within making, creating and traversing representations includes the following: the ways people talk, gesture, and sketch; the ways people create and move between sketches, computer aided drafting, vector-based drawings, and fabricated prototypes; reading oscilloscopes and other diagnostic equipment; reading measurements; and making notes to oneself.

Making, like engineering and design, is inherently a multimodal activity (Nathan et al., 2013). Formalized representations, like graphs and CAD drawings, play critical roles in the professional practices of engineers (Noss, Bakker, Hoyles, & Kent, 2007), just as informal sketching, drawing, and forms of expressive media are important supports for thinking and reflective action (Ferguson, 1992). In both making and engineering, the processes of constructing design embodiments—e.g., prototypes, products, personally meaningful artifacts—often begin with a sketch in a notebook, for example. The design evolves and takes form by re-drawing, mocking up the design in digital formats, or even producing formal design files on a computer. These multimodal texts become integral to the conversations that makers enter into with the design ideas, materials, and the possible processes for constructing the artifact (Schön, 1992).

Scholars have examined how learners move across different representational forms within certain kinds of activity. In mathematics, representational fluency has been described as mobility across conventional forms, linking and connecting them, producing and interpreting them, all for the purposes of coherent, disciplined mathematical reasoning (Kaput, 1991; Goldin, 1998; Nathan, Stephens, Masarik, Alibali, & Koedinger, 2002). In some ways, the very nature of mathematics is so entangled with the representational forms it takes that these conventional or recognized forms become not only ways of expressing and communicating mathematical ideas, but also mathematical objects in their own right (Sfard, 1991). It makes sense, then, that attention is paid largely to the conventional representations so foundational to mathematics practice: graphs, tables, and algebraic notation, for example. When moving between and among representations, fluency describes the depth of knowledge and familiarity with one particular system, and how individuals are able to establish and use connections between systems (Lesh & Doerr, 2003). The ability to move across these forms is considered an indication of one's knowledge and facility with particular mathematical concepts (Goldin & Shteingold, 2001). Others have built on these ideas of fluency in science, and engineering, focusing on the movements between forms within particular kinds of activities like engineering (Johri & Lohani, 2011; Johri, Roth, & Olds, 2013; Nathan et al., 2013) or modeling in science (Gravel, Scheuer, & Brizuela, 2013; Gravel & Wilkerson, 2017).

We seek to extend the conversation beyond mathematics, where many conventional forms are well-established, parsimonious, and elegant, into ways of working that involve multiple materials, ill-defined problems, and novel arrangements of tools and representations. We use fluidity to expand the notion of fluency within contexts that require new considerations of how individuals use and traverse representations, like making. Making spaces share the characteristics of multimodality and transdisciplinarity, among others, and this rich representational terrain is an opportunity to explore complex combinations of forms, systems, and activities to better understand how representations support creative production.

The idea of representational fluidity means more than being able to move from one representation to the other. It speaks to the polysemous nature of all representations, the ways in which different people interpret texts in different ways. It also speaks to the ways in which creators are often in conversation with the materials they are using, responding to the materials' properties and characteristics. More broadly, representational fluidity emphasizes the range of idiosyncrasies and formalizations present in the varied ways that individuals engage with representations to begin challenging the exclusive attention paid to conventional forms, or privileged languages (Bamburger, 1991) within STEM fields. In part, this effort expands what is considered valued in STEM (Bang, Warren, Rosebery, & Medin, 2012), going beyond only the conventional forms to include the invented, messy, and informal "in betweens" that many people use to navigate their work while they produce and share knowledge. In this chapter, by emphasizing the production, interpretation, and relationships between multiple representations in making, we illustrate how developing fluidity supports the maker's work, and in turn is a means of engaging in the literacy practices already valued in STEM.

We present cases from three different making spaces to begin articulating the features of fluidity. In particular, we examine the notion of traversals, or the fluid movement between and across representations of varying formality. Second, we explore how youth makers invent their own representations to help navigate and work within conventional forms, like computer code and electronics diagrams. And finally, we explore the specific role of "sketching across materials," how an emphasis on sketching can encourage sense making as makers "sketch" across different materials and representations. These cases paint a picture of the nuance and sophistication of fluidity, using traversals, inventions, and sketching as organizing ideas. As the maker engages in reflective conversations with materials and tools (Schön, 1992), representations become objects to think with (Kafai, 1995; Turkle, 2007). Positioned in this way, representational fluidity captures the practices of both using and producing representations in service of making.

Representational Fluidity Across Sites

We structure the chapter as a cross-case analysis of three sites, and how representations are treated in the work of the participants in these communities. At

Artisan's Asylum we describe how professional makers and hobbyists embraced the informal nature of their work, moving rapidly between representations, and how they purposely used sketches rather than more formal, technical drawings. We emphasize the central role of sketching, and informality, in how representations are supports in the work. At SETC, we describe how youth invented representations to support their traversals, from sketches, to schematics, to diagramming electrical circuits, to computer code, all in the process of fabricating an interactive water piano. Finally, at Nedlam's Workshop, we present how an emphasis on sketching as an approach to making, not simply a paper and pencil activity, supported new makers as they scope and frame their problems using different materials. We illustrate how this idea of "sketching across materials" is a sense-making activity, similar to the ways the experienced makers at Artisan's used sketching as a thinking tool in their everyday practice.

The cases were constructed following a comprehensive coding of all data (described in Appendix A). All instances of the practice "representation" were selected and analyzed using a content analysis approach to yield themes (Charmaz, 2014). These themes were used to revisit the data to construct specific cases of the ways representations were used in the three making spaces. One of the cases draws from a design we instituted to further explore the role of sketching in making, which became evident to us at Artisan's Asylum. In this design case, we articulate how we introduced particular design features to support learners in developing their project, and in developing the representations that structured their work. The cases in this chapter each highlight a component of representational fluidity—traversals, inventions, and sketching—as we assemble a description of the creative, multimodal, and multifaceted ways in which makers use representations in their work.

Traversals at Artisan's Asylum

There were three main reasons people created representations at Artisan's Asylum: to brainstorm or work out ideas in ideation, design, or iteration processes; to provide specifications for fabrication; and to share work with others. Ideation and design tended to involve visually sketching ideas and objects with pencil or pen and paper (see Figure 5.1). Almost everyone we talked to, both those who identified more as artists and those who identified more as engineers, kept some sort of notebook for their ideas.

Representations are sources of information that are enabled and supported by a comfort and familiarity, or fluency, with many different representational systems. Representational traversals were part of the everyday work of the professionals at the Asylum. The various forms were more than just representations of particular ideas, they were tools to think with and through. As one of the experienced makers in our initial interviews told us,

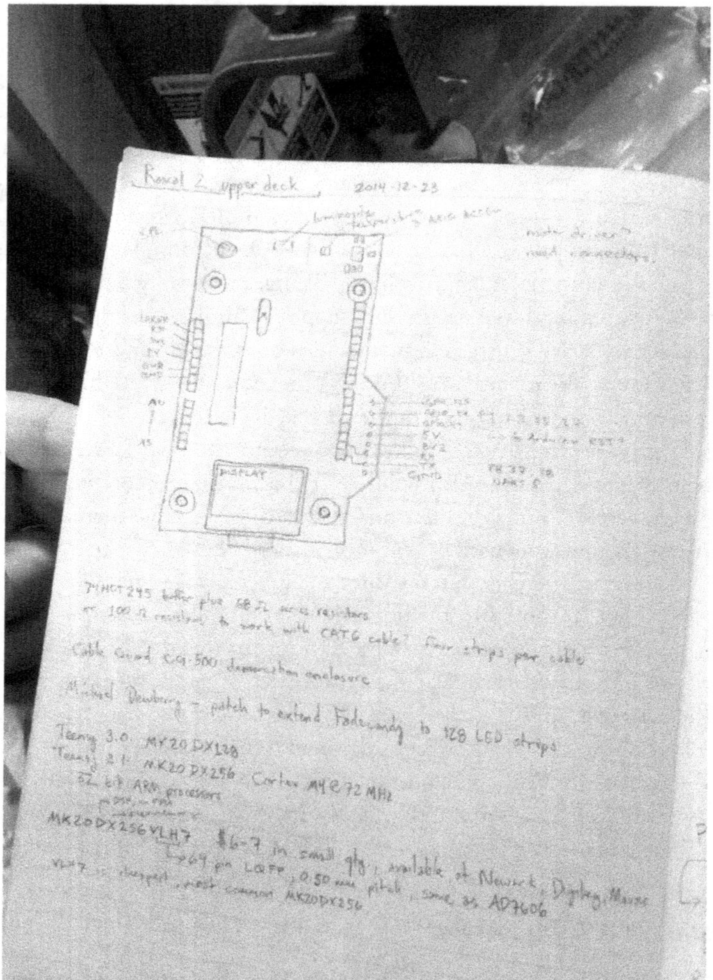

FIGURE 5.1 A page from Brandon's design notebook.

Over time it just became part of my practice when it comes to sketching things and developing ideas. Half the time I use a pencil and half the time I use pieces of wood and scraps and paper and whatever else. It's just a way of working out ideas in the physical world.

Walter, Artist

Makers at Artisan's shared many of the same representational tools as professional engineers, especially those that prepared files to manufacture on laser cutters, computer numeric controlled routers (CNC), and 3D printers. The main Computer Aided Design (CAD) software we observed was Solidworks, a proprietary software used widely in industry. Some makers also embraced open-source values and preferred open-source mechanical CAD and electrical CAD software for designing those kinds of components for their projects. One of the ways in which making at Artisan's differed from working at an engineering firm was that the makers often moved quickly to manufacturing after sketches and some rough planning in computer-aided software. A former engineer, Richard, told us:

> I used to run an engineering department, and we used to be very disciplined in having drawings, and signing off on drawings—it could take a really long time to do anything . . . in this kind of environment, you can work a lot more quickly. So, the representations are often not solidified in the ways a more traditional engineering system might work. If it ever gets to a point where this thing needs to be made in the thousands, or tens of thousands, then it needs to be tightened up. In this kind of environment, you can work much more quickly. The thing you make is the most important thing, the representations are an aid, but they don't become so solidified that they slow the process down.
>
> *Richard, Engineer*

Richard illustrated an important distinction between the ways representations are created, shared, and used in formal processes of engineering firms, and what making as an activity provides: flexibility and a kind of fluidity. Experienced makers, often trained within particular disciplines, created layers of personal and cultural meaning through the representations they used (Richard was formerly an electrical engineer). The conventional forms they produced and used, like CAD renderings, were powerful tools to think with in part because these trained makers had previously built practices for using the conventions and grammars of these systems to support their work. However, within making, the adherence to conventional rules for working with these valued forms (i.e., those that are linked to STEM conventions such as coding languages, histograms, or electrical symbols) is relaxed, and even challenged by some. Treating the representations one produces as thinking tools, rather than formal records of the work makes those representations more fluid, both in form and in practice. For example, we observed two engineers at Artisan's evaluating a design prototype they 3D printed. As they looked at the curvatures, clearance, and how the shapes for the part came together, they had ideas for iterating to improve the design. Where engineers at a firm may mark these iterations by a revision to the CAD file, so as to document decisions, these two took a Sharpie marker and began to mark up the actual prototype.

> They both visualize what the product "will" look like. Chris draws on the actual prototype to show what it might look like and where it will go. Chris uses his phone to take photographs of the different prototypes they were talking about.
>
> *(Fieldnotes)*

Sketching on paper, objects, and screens extends thinking into the material space, where ideas and materials are brought into conversation with each other. At Artisan's Asylum, and for some other experienced makers we interviewed, sketching was a central practice, as central to their work as knife skills to a chef, or measurement to a carpenter; a foundational tool and a stance on their work—sketching was the way one thinks with objects, representations of them, or the materials themselves. We take care to specifically address the role of materials here, and how sketching across materials is a way to understand materiality, and the conversations makers have with the things they manipulate, fabricate, and share. Karen, a jeweler, highlighted an important aspect of this in her interview:

> My worry sometimes, when you're sitting in front of a computer, and all your design is there, that you're so bound in how the thing is kind of looking. . . . You lose your literacy. . . . And, I find that kind of going back to the hand, and going back to the sort of old way to work invites more happy accidents.
>
> *Karen, Jeweler*

Making is a learning activity, a means of exploring new spaces and ideas. Karen acknowledged how different representational tools—e.g., computer tools—may allow for the kinds of sketching and the fluid uses of representations that Richard described; however, materials are also an important medium in the work of "sketching" an idea. And for Karen, shedding the computer, or even paper and pencil, and working directly with the materials took on the essence of sketching, where ideas emerged and "happy accidents" led to new directions for the work. Makers traversed representations and materials with this fluidity, allowing them to occupy spaces between conventional forms, often with idiosyncratic, personal, and invented "conversations" with the materials.

> [. . .] the sort of improvisatory process happened and the sketching, building, call-and-response, kind of a conversation with the materials I had at hand sort of played out and then the instrument got finished a few days later. . . . At that point I generally don't have a fixed sense of what it's going to become at the end, I just have a notion and then I introduce that notion to the materials and I really believe in this idea of it being a conversation.
>
> *Walter, Artist*

Walter's quote highlighted the object, or material, as representational thinking partner. The act of sketching is just that, a thinking activity. The pencil and paper are extensions of the maker's imagination, sketching is a way of envisioning, dimensioning, and bringing to life ideas and arrangements; it is a form of thinking for the designer (Ferguson, 1992). With fluidity, we are projecting that notion of sketching as thinking into multimodal spaces, where materials, representations, and objects comprise the collection of tools that support the maker's work.

Our observations revealed that representations are everywhere at Artisan's Asylum, which is expected. The movements between them, and the modes of traversing them are best captured by the notion of fluidity: a loose, personal path through ideas, where value is distributed across forms and formalities. While the makers at Artisan's were often well versed in the conventional and formalized representations of their chosen domains, their practices also blurred the lines between formalizations through using and inventing representations for sketching and exploring, even while in the fabrication process. The importance of inventions in making is further highlighted by how youth makers used representations at SETC.

Inventions at South End Technology Center (SETC)

At the South End Technology Center's summer *Learn to Teach, Teach to Learn (L2TT2L)* program, representations were also everywhere. However, we did not observe nearly as much "paper and pencil" sketching in the youth's work as we did at Artisan's Asylum. The few examples that we did record were very similar to what we saw at Artisan's: youth sketched ideas to discuss them with others, or they sketched on a piece of paper before translating that to something more formalized or conventional, like CAD software. While sketching and traversals were themes at Artisan's, at SETC we saw sketching serving another kind of role, it became a space for inventions—sketches intended for very specific purposes that did not necessarily adhere to the rules of some conventional system—that helped makers map different representations onto each other. We focus on one group to explore the role of invention in representational fluidity.

One project was called "WaterWorks." Four youth collaborated to design and fabricate an interactive light and sound water piano. Their vision was that a user would place a finger in a vessel of water to play a musical note and produce a corresponding color on a multicolor LED. This project involved multiple electronic components that all needed to operate in synchronization with each other; it was a complex and creative design, featuring multimodality in the user's experience and in how they needed to go about fabricating the object.

Within the physical computing world, "pins" are used to direct microprocessors, like an Arduino, to detect certain signals related to certain components. For example, commands in the programming code might tell the Arduino to "turn on" the output (e.g., a light) attached to "pin 7" if a certain signal is received. WaterWorks's project involved many different LEDs, different containers of water, the Makey Makey, and the Arduino One can imagine the entanglement of wires going from pin to pin on the various devices (see Figure 5.3) as they built their project. This was confusing and cumbersome to manage.

> The wiring kind of got messed up a little bit at one point, so I had to take it all off and rewire the whole thing, so I wrote it down just in case I had to do it again . . . I wrote down what numbers I put the pins at, and what color I put it in the code.
>
> *Mary, member of WaterWorks*

As Gus and Mary began programming and building their project, one of the mentors recommended they create a "map" linking pins to each function. Mary shared how this came about, that the wiring got "messed up" at one point, and they needed a way to keep track of all the various connections. The first note they wrote (Figure 5.2) helped them keep track of what wires went where. But as their project gained complexity and new components were added, the notes that they jotted down quickly required revision. Together with their mentor, they created another map that included each pin-to-pin correspondence, as well as a diagram of how the components would be wired together, which included conventional circuit diagramming symbols (e.g., resistors, power sources). This second iteration of the map (Figure 5.2) was carried by the group every day as they continued to develop their project. Notes were added to it. Water was spilled on it. And a photocopy was made so that they had a backup. "Where's that paper?" was commonly heard as they started work each day.

This "map" (shown together with a second version, their electronic components, and the on-screen Arduino code that was used to control the project, Figures 5.2 and 5.3), became the scaffold for their work developing the electronic and computational components of the project. It supported them, along with other resources, as they navigated coding the Arduino, which was a new form of STEM activity for them.

> I just started on the coding team. . . . I wasn't really sure how to code. But, I didn't know what exactly they were doing, and what vocabulary they were using. . . . I couldn't be able to write it myself, starting from scratch, even blink an LED, but if I had a resource I could go to. . . . I could read a code and understand what's happening.
>
> *Sanna, member of WaterWorks*

Sanna talked about the map, and working from example segments of code, as aides that allowed her to progress from someone who had never coded to someone who contributed significantly to the development of this project and to

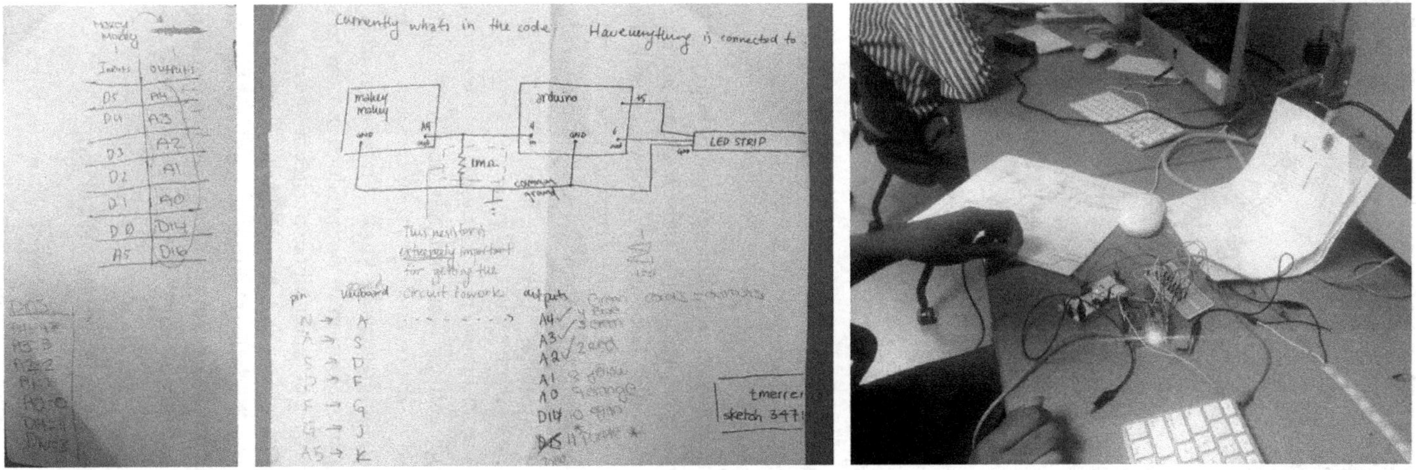

FIGURE 5.2 The evolution of WaterWorks's invented schematics to support their work coding and wiring their project, including their first note (left), the elaborated version of their map (center), and both being used to troubleshoot connections (right).

FIGURE 5.3 The invented coding and pin maps for WaterWorks, alongside the Arduino code controlling their project.

complex computational tasks like troubleshooting and debugging. Creating and using the maps were literacy practices. Sanna and her group members used their invented representations to troubleshoot an issue that emerged regarding which LEDs were firing and where possible wiring errors may be on their breadboard (Tucker-Raymond, Gravel, Klimczak, Wagh, & Ren, in press). Using a multimeter to examine each connection, the map helped structure their debugging process until they were able to resolve the LED issue.

Ultimately, the representations they created bridged their ways of understanding and representing the project to the conventional nature of the coding languages and infrastructures required by the technologies they chose to use. Invented representations developed to achieve particular kinds of tasks are further evidence of fluidity, informality, and personalization in the production of representations in making. These inventions were authentically generated as sense making devices, as tools to think with, and as vehicles that supported the traversal of both familiar and unfamiliar representational worlds. Coming to see sketching, and inventing new representations, as aides for exploring new ideas and technologies embodies the idea that making spaces foster literacies that support STEM learning.

Sketching Across Materials at Nedlam's Workshop

The final case we present is of a making course held in Nedlam's Workshop, the making space at our partner school. We supported the teacher, Ms. Freewoman, in the development of an elective course called *Creative Design and Engineering.* The course foregrounded literacies in making and was designed to help students learn new kinds of making processes and skills as they designed and made things to solve problems in their lives and communities. This course became a venue to implement designs that allowed us to both support students' STEM literacies in making and to study the ways these literacies developed.

The course was designed with a heavy emphasis on sketching and creating representations. Each student received a notebook, and throughout the making space we distributed clipboards with white paper and black felt-tipped pens. Each day students were asked to write, draw, sketch, or keep records in their notebooks. Our conjecture was that foregrounding sketching in all facets of the course would encourage students to adopt the stance that "sketching is thinking," similar to what we observed at Artisan's Asylum. The heavy encouragement of sketching began to appear in how the students worked and talked with each other. "Show me," was a common response when students asked questions of each other or when they asked facilitators for advice or help. Conversations were often mediated by a sketch on a clipboard, a drawing on a notebook page, or even doodles on the workbench surface if that was convenient. We share the case of two students, Antonio and Rubens, whose design journey traversed different materials and representations, where each iteration became a means for furthering their inquiry into the problem and their understandings of the challenges and solutions.

"*They're cold, there's no place to charge your phone, and they smell like pee.*" Students in *Creative Design and Engineering* agreed that taking the public bus to and from school in the winter in New England was not a pleasant experience. Together as a class, they scoped this problem, and one of the core components that they selected for this design was a heated floor. Rubens and Antonio chose to work on this issue, and in conversations with us and Ms. Freewoman the idea of radiant flooring was raised. Antonio and Rubens liked this idea, and they set about trying to find information online about these systems. Unfortunately, most of what they found was intended for contractors or engineers, and the information was not particularly helpful in moving them forward. Unsure of what to do next or where to turn for help getting started, Antonio and Rubens were rather stuck.

We noticed this issue emerge across many of the groups working on the bus stop project. Their goals of building an improved bus stop were very real to them, but it was a big, complex, and multifaceted project. No matter the specific problem—how to power the bus stop, how to heat it, how to get public support for the project—the students had wonderful ideas, but they struggled to move out of the phase of discussion and into action. Together with Ms. Freewoman, we designed a set of small "playgrounds" corresponding to each of the problems the groups were considering. This design was intended to support the students' continued exploration into the ideas they developed, but at smaller scales and with a simpler set of materials. Narrowing the challenge, we felt, would help them better scope the overall dimensions of the problems they faced. One group was working on providing power for the bus stop, and their "playground" consisted of small solar panels and a challenge to see if the students could use the panels to charge a cell phone. The group charged with building the structure for the bus stop made prototypes of the frame with popsicle sticks and other small pieces of wood. This helped them imagine the overall dimensions so they could begin cutting wood to build the life-size prototype. Rubens and Antonio, the heat group, were tasked with building a model of how they would design the radiant heat tubing to go underneath the stop. We wanted them to ask: how would the tubing be laid out? How would it be held in place? What would it be made of?

Rubens began with a sketch to help his team think about the problem. His drawing captured what he imagined the tube design to look like (Figure 5.4). Ms. Freewoman introduced the pair to Easel, software to design and cut material using the CNC mill, and they moved from their sketch of the possible solution to designing a platform they could use to test different tube configurations. Antonio said they, "Got a picture from Google" and used this as a template for their design in Easel (Figure 5.5). As they dimensioned the Easel file, they measured the tubing diameter and made estimations about the size and shape of the "dots" on the wooden frame (Figure 5.6). Ultimately, the result was a design file for the CNC mill that could be used to cut a template, on which they could explore the tubing configurations (Figure 5.7).

FIGURE 5.4 Rubens's first sketch of how to heat the bus stop floor using power from solar panels.

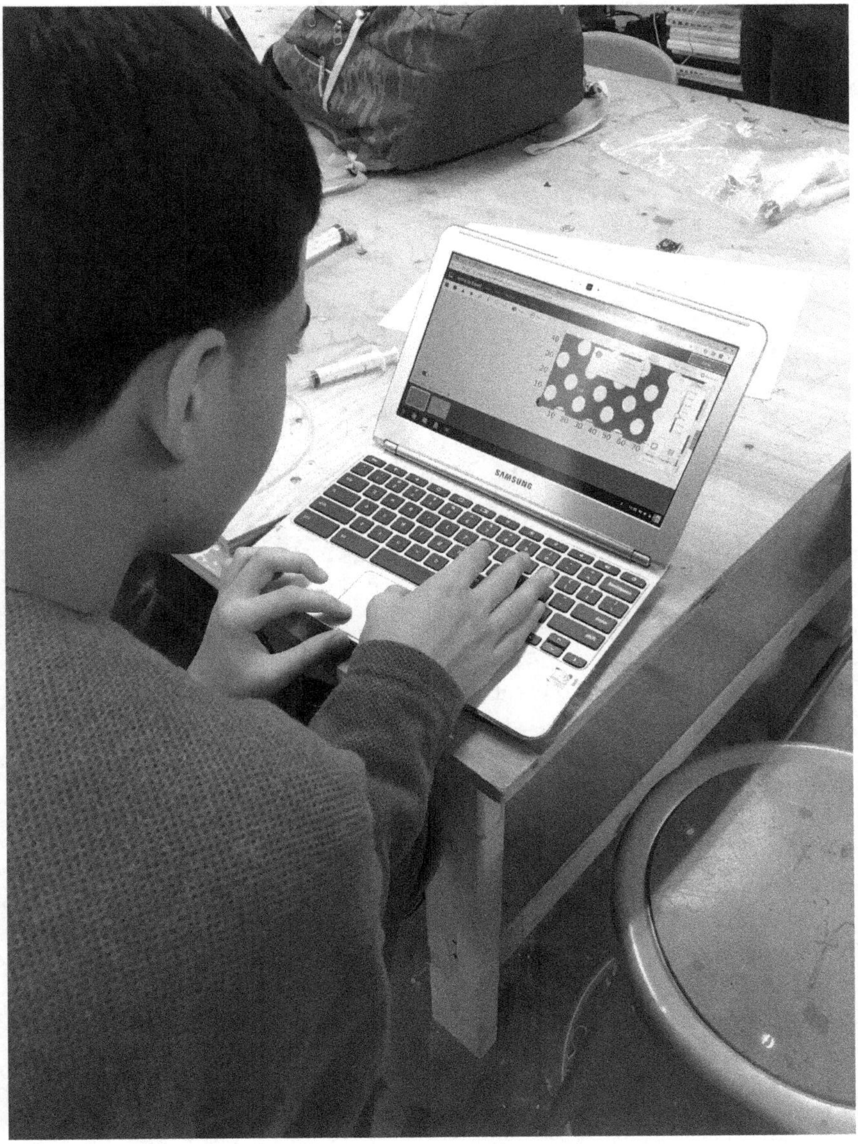

FIGURE 5.5 Antonio constructing the cut file in Easel to build the prototype using the CNC mill.

Rubens and Antonio cut their design out of wood on the CNC Mill and at over five hours to produce, "It took a very, very long time," Rubens said. With the template constructed, they explored different configurations, eventually gluing their final design in place. They used syringes to push water through the tube to model the liquid in a radiant flooring system. Their cut design served

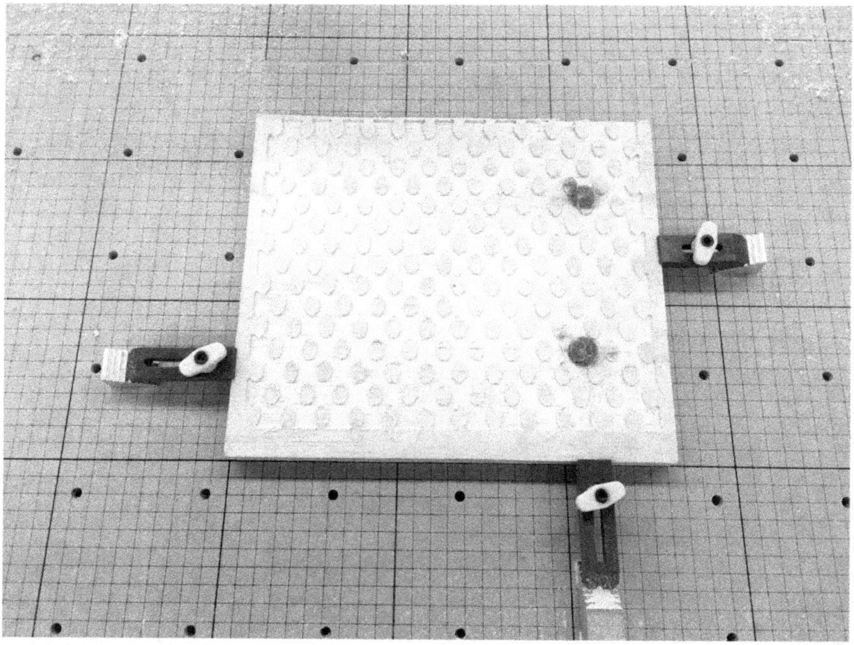

FIGURE 5.6 The wooden prototype made using the CNC mill.

as a representation of the problem, a dynamic one in the sense that they could manipulate and reconfigure it to explore options and to inform their decisions.

Their idea, which started as a sketch on paper, was translated to a vectorized image on the computer, and was now a wooden object that became a "sketch-pad" for their thinking about the arrangements of the tubing for the radiant floor heating design. They wove the tubing in different ways to explore possibilities, before gluing their final decision in place. We understand this fluid traversal as sketching across materials, where their ideas about the problem were coupled with the particular representations they produced as they developed the prototype. "It's like a map," Rubens said about their wooden platform. Later in the project, we asked them to reflect on their process:

RUBENS: At the time, we didn't know how big the bus stop was going to be. Depending on the structure team. We took the floor, and traced it, to start to get an idea.

ANTONIO: We wanted to be more flexible. Our first idea was just to trace something. This gave us more freedom to put like this (one way), or this (another way).

Rubens and Antonio spoke of their prototype as a "map" that helped them understand the problem. Their design relied on the decisions of other groups,

FIGURE 5.7 Antonio and Rubens trying out different tubing configurations on their "map."

as they planned to embed it in the frame and power it using the solar panels on the roof, solutions two other groups were still developing. As a result, they built something that would allow them to continue to understand and work on their problem, but that was "more flexible," and provided "freedom" to explore different configurations. As they played with their design, they began to ask questions: *"What's going to heat the water in the tube?"; "How do you generate heat?"; "Wait, though, we don't want it to heat up when it's hot out. How do we turn it on and off?"*

The questions that emerged for them, and the work they did exploring these questions with different materials (e.g., using heat tape to see how they might generate heat), amounted to a continually refined sense of the problem. Their traversal, from sketch to computer aided design, to physical prototype, was a fluid movement between representations that helped them understand the problem. They "sketched" representations to think with, across different materials, that helped them see new aspects of the problem, and to rephrase questions in ways that moved them forward (e.g., how do you generate heat?).

Rubens and Antonio finally built a "mock up" of their design (layering tubing through the floor of the full-size bus stop frame), and used it to share a story of their design process and thinking at the final exposition of the bus stop project (Wilkerson & Gravel, in press). That reflective act further illustrated the ways they not only progressed through their design, but also the ways they contended with rich disciplinary content and questions:

RUBENS: For heating the water inside the pipes, we'd have to calculate the volts and the volume of water so it would get hot enough.
ANTONIO: And the boiling point . . .
BRIAN: But you were saying you thought there may be things besides water?
ANTONIO: Yeah, like oil probably, that had a higher boiling point. To get it probably to a higher temperature.

The threads of Ruben's and Antonio's work developing and sketching the solution for heating the bus stop wove across different forms—paper, digital design software, fabricated objects—as they refined their understandings of the problem. Similar to the experienced makers, the youth sketched the problem across materials to map the space of a design, using different representations to understand what a radiant floor heating system entailed. The digital design file led them to wonder about the dimensions of the tubing, and the wooden template let them to think about the materials and sources of heat. Overall, movements across different systems and materials helped the students make sense of their work.

From Fluency to Fluidity

The three cases in this chapter illuminate different aspects of what we have called representational fluidity. Our goal is to expand on existing conversations about multiple representations, and the movement between representations, in STEM learning (e.g., Lesh & Doerr, 2003; Kaput, 1991; Gravel & Wilkerson, 2017). Each of the cases described elements of how makers consume and produce representations in their work. They also address the coexistence of multiple forms situated within the contexts of the making spaces where they work. All three cases showed how making can blur lines between conventional forms of STEM representations, and centers the use of invented, informal, and idiosyncratic forms.

Walter painted a beautiful image of how making brings different representations and materials together into conversation with the maker, as he described an "improvisatory process happened and the sketching, building, call-and-response, kind of a conversation with the materials I had at hand sort of played out." Richard also noted how informality supports rapid iteration, in part by abandoning strict adherences to representational procedures. Free from the drawing protocols of an engineering firm, he felt he could move quickly and creatively with his ideas. For all of the makers that we interviewed and observed, the loose and personal dialogues with representations allowed them to become objects that helped makers scope, frame, understand, and explore the nature of their projects. The rapid traversals across the different forms were characteristic of how representations shaped and supported the work of makers.

Bringing the lens of fluidity to a making space, we can examine how representations function within a community or group with established norms and values. The Youth Teachers at SETC, supported by mentors, constructed representations together to understand their problem. They literally mapped the components and generated shared understandings of the work they needed to do. The maps were iterative inventions that became foundational to guiding their future work. Inventing ways to navigate conventional forms—e.g., Arduino code, circuit diagrams—allowed them to move between systems and objects as they figured out their project. WaterWorks is decidedly a story about learning through making, and the invented representations were how the youth makers learned and produced knowledge in STEM.

Finally, we extend the purpose of "sketching" that was so prevalent at Artisan's to the production of representational objects in a variety of modalities. Sketching across materials served to organize successful cycles of discovery and problem solving for youth makers. As Antonio and Rubens moved from paper and pencil, to a computer design file, to a digitally fabricated object, they produced representations of the problem space. Yes, these were objects too, but understood through the lens of sketching across materials, each round of production served a similar mediating role as sketching did for the experienced makers. The "map" they produced on the CNC was a way to understand the problem as well as explore the STEM ideas and content that emerged in their designed solution.

The idea that producing representations is useful, even integral, to STEM learning is not new (e.g., Lehrer & Schauble, 2002). diSessa and Sherin (2000) outlined the notion of meta-representational competence to highlight that individuals can, and do, produce, refine, and learn new representations of the physical and even imagined situations they encounter. While "competence" is a useful construct in establishing that youth are producers of representations, we see fluidity as a way to extend that notion into multidisciplinary and social spaces for two primary reasons. First, we wish to highlight the range of representational formality that exists in making, and how the traversals across and through forms blur lines between convention and invention. diSessa and colleagues also highlighted this point, suggesting that youth can invent approximations of conventional forms, and that their invented forms are ways into understanding and becoming conversant with systems like Cartesian graphs (diSessa, Hammer, Sherin, & Kolpakowski, 1991). An emphasis on the messy, personal, and informal representations in making helps to complicate notions of "fluency" (Nathan, Stephens, Masarik, Alibali, & Koedinger, 2002), recognizing that conventions are only conventions because they have proven useful over long periods of time by critical masses of individuals. However, the blurring of conventions and inventions foregrounds the ways representations aide thinking, becoming partners in the iterative refinement of ideas and designs in making.

Second, fluidity highlights how relaxing the rules of a system makes space for seeing the production of representations as the collective production of knowledge, highlighting the multimodal and multidisciplinary values of making spaces. The stories of WaterWorks and Antonio and Rubens both demonstrate how their inventions, or successive sketches across materials, became ways for them to build knowledge together. Relocating the conversation from individual competency to the shared space of a community learning together positions fluidity as a phenomenon that spaces can engender. Fluidity may describe the individual maker's movements with representations, but it can also describe how groups of people approach traversals and inventions together. Whereby collaborative production of invented forms in support of making, like the youth cases in this chapter, encourages movement across different representations, inventions and conventions.

Ultimately, we think fluidity is of particular importance because of the ways power operates through representations in STEM fields. Access to dominant discourses—including the conventional and chosen forms of knowledge production and sharing, like Cartesian graphs or C++ syntax—is limited when these discourses only include historically accepted forms of the disciplines and do not leave room for hybridity, informality, or play. One's familiarity, and even comfort, in engaging with established representational forms can stratify participation in STEM. We would argue that making spaces, where representational fluidity is practiced, promoted, and celebrated, expand the points of entry to these privileged forms that often serve as gatekeepers to STEM participation. A space that encourages inventions—maps of Arduino pins, or Sharpies drawn on 3D printed prototypes—promotes multiple ways of knowing and producing knowledge. Acknowledging these power differentials in how making spaces are built and operated can help in the effort to build spaces that are more equitable and that promote the participation of people from groups who have been marginalized in STEM.

References

Bamburger, J. (1991). The laboratory for making things. In D. A. Schön (Ed.), *The reflective turn: Case studies in and on educational practice* (pp. 37–62). New York, NY: Teachers College Press.

Bang, M., Warren, B., Rosebery, A. S., & Medin, D. (2012). Desettling expectations in science education. *Human Development, 55*(5–6), 302–318.

Charmaz, K. (2014). *Constructing grounded theory.* Thousand Oakes, CA: Sage Publishing.

diSessa, A. A., Hammer, D., Sherin, B., & Kolpakowski, T. (1991). Inventing graphing: Meta-representational expertise in children. *Journal of Mathematical Behavior, 10*(2), 117–160.

diSessa, A. A., & Sherin, B. L. (2000). Meta-representation: An introduction. *The Journal of Mathematical Behavior, 19*(4), 385–398.

Ferguson, E. S. (1992). *Engineering and the mind's eye.* Cambridge, MA: MIT Press.

Goldin, G. A. (1998). Representational systems, learning, and problem solving in mathematics. *The Journal of Mathematical Behavior, 17*(2), 137–165.

Goldin, G. A., & Shteingold, N. (2001). Systems of representations and the development of mathematical concepts. In A. A. Cuoco & F. R. Curcio (Eds.), *The roles of representation in school mathematics* (pp. 1–23). Reston, VA: NCTM.

Gravel, B. E., Scheuer, N., & Brizuela, B. M. (2013). Using representations to reason about air and particles. In B. M. Brizuela & B. E. Gravel (Eds.), *"Show me what you know" exploring students representations across STEM disciplines* (pp. 163–182). New York, NY: Teachers College Press.

Gravel, B. E., & Wilkerson, M. H. (2017). Integrating computational artifacts into the multi-representational toolkit of physics education. In D. F. Treagust, R. Duit, & H. E. Fischer (Eds.), *Multiple representations in physics education* (pp. 47–70). Dordrecht: Springer.

Johri, A., & Lohani, V. K. (2011). Framework for improving engineering representational literacy by using pen-based computing. *International Journal of Engineering Education, 27*(5), 958.

Johri, A., Roth, W. M., & Olds, B. M. (2013). The role of representations in engineering practices: Taking a turn towards inscriptions. *Journal of Engineering Education, 102*(1), 2–19.

Kafai, Y. (1995). *Minds in play: Computer game design as a context for children's learning.* Hillsdale, NJ: Lawrence Erlbaum Associates, Inc.

Kaput, J. J. (1991). Notations and representations as mediators of constructive processes. In *Radical constructivism in mathematics education* (pp. 53–74). Dordrecht: Springer.

Lehrer, R., & Schauble, L. (2002). Symbolic communication in mathematics and science: Coconstituting inscription and thought. In E. D. Amsel & J. Byrnes (Eds.), *Language, literacy, and cognitive development: The development and consequences of symbolic communication* (pp. 167–192). Mahwah, NJ: Lawrence Erlbaum Associates.

Lesh, R. A., & Doerr, H. M. (2003). *Beyond constructivism: Models and modeling perspectives on mathematics problem solving, learning, and teaching.* New York, NY: Routledge.

Noss, R., Bakker, A., Hoyles, C., & Kent, P. (2007). Situating graphs as workplace knowledge. *Educational Studies in Mathematics, 65*(3), 367–384.

Nathan, M. J., Srisurichan, R., Walkington, C., Wolfgram, M., Williams, C., & Alibali, M. W. (2013). Building cohesion across representations: A mechanism for STEM integration. *Journal of Engineering Education, 102*(1), 77–116.

Nathan, M. J., Stephens, A. C., Masarik, D. K., Alibali, M. W., & Koedinger, K. R. (2002). Representational fluency in middle school: A classroom study. In *Proceedings of the twenty-fourth annual meeting of the North American chapter of the International Group for the Psychology of Mathematics Education* (Vol. 1, pp. 462–472). Columbus, OH: ERIC Clearinghouse for Science, Mathematics and Environmental Education.

Schön, D. A. (1992). Designing as reflective conversation with the materials of a design situation. *Research in Engineering Design, 3*(3), 131–147.

Sfard, A. (1991). On the dual nature of mathematical conceptions: Reflections on processes and objects as different sides of the same coin. *Educational Studies in Mathematics, 22*(1), 1–36.

Tucker-Raymond, E., Gravel, B., Klimczak, S., Wagh, A., & Ren, A. (In Press). "It will be a Symphony of Children": Learning STEM literacies through technologies of the heart. In A. Wilson-Lopez, E. Tucker-Raymond, A. Esquinca, & J. A. Mejía (Eds.), *The literacies of design: Studies of equity and imagination in engineering and making.* West Lafayette, IN: Purdue University Press.

Turkle, S. E. (2007). *Evocative objects: Things we think with.* Cambridge, MA: MIT Press.

Wilkerson, M. H., & Gravel, B. E. (In Press). From personal artifacts to shared solutions: Storytelling as a support for Constructionist activity. In N. Holbert, M. Berland, & Y. Kafai (Eds.), *Constructionism in Context.* Cambridge, MA: MIT Press.

6

ONLINE SHARING GENRES

How-Tos, Crowdsourcing, and Blogging

> **Focus Literacies in This Chapter**
>
> **Sharing finished or almost finished artifacts**: Demonstrating made artifacts, producing web pages, or using social media to promote work.
>
> **Documenting making processes and/or milestones**: Photographing, writing about, or recording the steps, processes, and output of a project for oneself or as a reference for others.

This chapter describes the ways in which makers share their work online. Although online texts live on screens and are not always physically located in making spaces, they are created by makers either in the process of making or to share something they have made. The chapter focuses on presenting three different ways of sharing from our participants: a) one youth creating a "how-to" to make a living hinge on the website Instructables.com; b) two adult entrepreneurs using Kickstarter to crowdsource financing for their product, a DNA replicator; and c) two blogs, one created by youth in a summer program—to document and share their work (a water piano) as inspiration for future program youth—and, a second one from a cosplay (i.e., costume play) maker who blogged about her process of making a helmet from a mold. We then offer recommendations for integrating means of authentic, public sharing in making spaces with learners.

First, as outlined in Chapter 1, literacies are social practices (e.g., Barton & Hamilton, 2012; Heath, 1983). That is, they connect particular literacy events (e.g., writing a crowdfunding request) to broader patterns of activity and ideology (e.g., entrepreneurship) in ways that are mediated by local values and purposes (e.g., biologists providing access to low-cost tools for biology). However, the social character of online literacy practices is even more evident. Online texts are written and prepared for an audience and that audience is expected to interact with the texts (e.g., buy a product, leave a comment on a blog, or make an object). In that sense, online texts are also performative of identities because to interact with different people for particular purposes makers must present themselves and their products in certain ways. In the very least, makers position their identities as producers that contribute materially to their world. As one of our adult participants from Artisan's Asylum, Ecco, stated, "I'm different from other people because I put things into the world and they consume them."

Sharing Genres

The constellations of literacy events, ideologies, local values and purposes, and the identities of the producers work together to create types of recognizable texts, or genres. In learning contexts, understanding genres helps learners to know what texts do and how they do it. During production of texts, knowing what texts do allows learners opportunities to try out ways of positioning their own identities. It also gives them opportunities to receive and respond to feedback from several sources. There are several genres of online sharing that are central to making.

Online Genres in Making

How-Tos/Tutorials (e.g., Instructables, YouTube video)
Crowdsourced Fundraising Campaigns (e.g., Kickstarter)
Blogging
Business Websites/Online Stores (e.g., etsy)
Personal Websites
Social Networking (e.g., Snapchat, Twitter, Instagram)

Although youth often engage in social networking sites on their own, we focus on the first three in the list above—how-tos and tutorials, crowdsourcing, and blogging because they reflect the kinds of academic writing students are asked to do in schools. How-tos are a common format in elementary school and precede lab reports that older students are asked to write when they record their trials and data systematically (NGSS Lead States, 2013). We describe one kind of how-to—written instructions with photographs—and the ways in which how-tos can be opportunities for identity play and connecting to a larger network of makers. Crowdsourcing requests are persuasive narratives from makers literally trying to make people "buy-in" to their idea. Finally, blogs in the making context can invoke narratives, how-tos, and more expository texts. They are often a mixture of all three.

Genre: How-To

How-tos, such as those found on Instructables.com posts as well as online videos, such as those on YouTube, are currently popular genres for makers to present how they made or fixed something or how they completed a process (Swarts, 2012). Instructables.com is a do-it-yourself (DIY) website on which people who make things leave instructions for how to make anything from recipes for food to jewelry to giant chairs shaped like baseballs. Whereas Instructables posts are considered free and open source to the community, YouTube videos are often made by people looking to recruit enough subscribers to earn money (Bhatia, 2018). At the same time, YouTube "has become a resource for young people both as a platform for expression, but also as a platform for informal learning" (Bhatia, 2018, p. 107). There are people who post on Instructables and have YouTube channels, websites, and Etsy stores at the same time. Given that there are so many corners of the Internet, finding the one that reaches one's audience is important. Despite being useful for both sharing what one has done and learning new information, YouTube is banned, or filtered from school computers, in many districts across the United States because content and comments are only lightly censored. We would advocate the removal of these kinds of bans to allow learners access to expertise that school teachers may not possess.

When communicating expertise online, posters or vloggers who present material are also establishing their identities (Tucker-Raymond, Gravel, Wagh, & Wilson, 2016). They draw on a range of resources including discursive competence and disciplinary knowledge and practice that, ideally, are simultaneously connected both to professional and lay practice (Bhatia, 2018). That is, people who post quality how-to videos have expertise in the topic and they know how to communicate that knowledge in a way others can understand.

While we were observing the South End Tech Center (SETC), one of the young people posted an Instructable about a wooden book cover with a living hinge that he had made using design software and a laser cutter (see Figure 6.1). A living hinge is composed of one continuous piece of material rather than two or three interacting parts, such as in a typical door hinge. In the first two months that the project was online, his post received over 4,000 page views and over 100 favorites. Naeem's work, and the way he shared it, clearly caught the attention of the makers that frequented the Instructables.com online community.

Naeem's multimodal digital composition in an online community (Figures 6.1–6.4) reflected his facility with digital literacies through the lens of

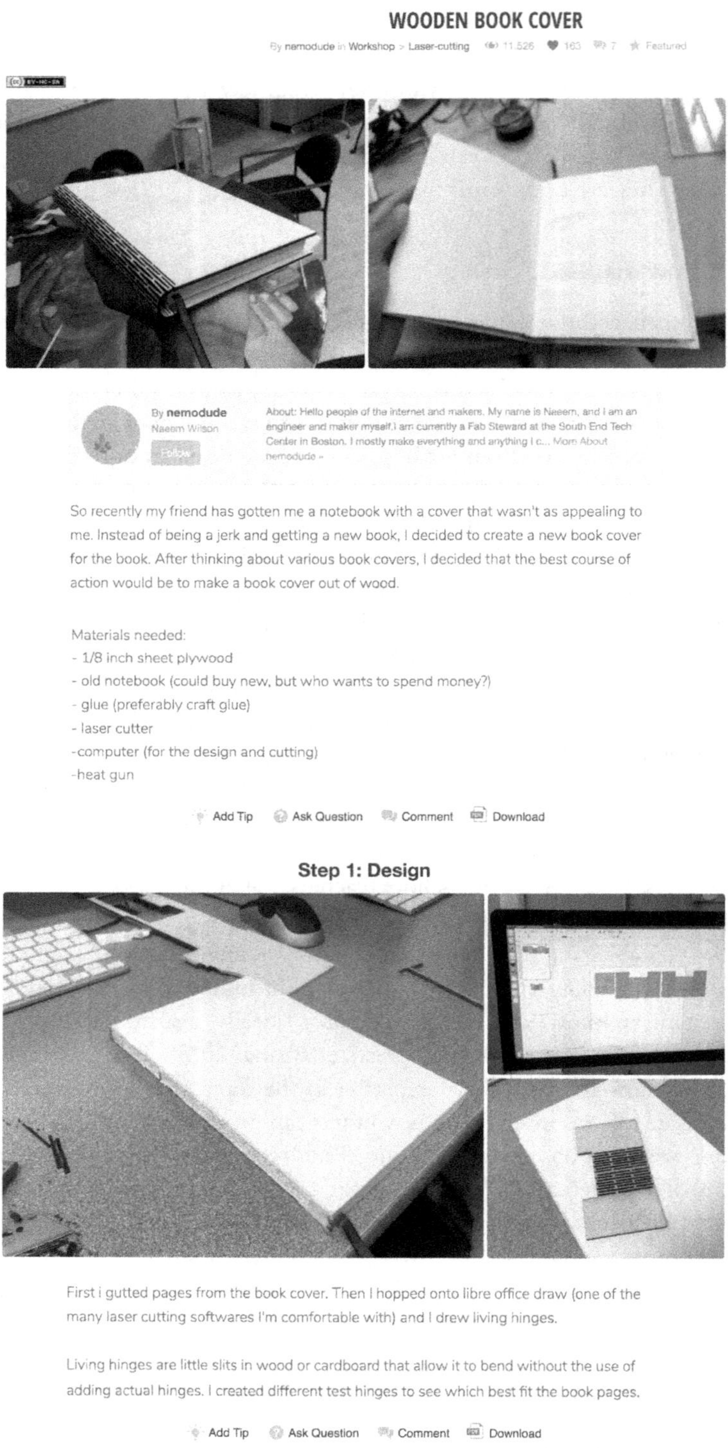

WOODEN BOOK COVER

By nemodude in Workshop > Laser-cutting 11,526 163 7 Featured

(cc) BY-NC-SA

By **nemodude**
Naeem Wilson
Follow

About: Hello people of the internet and makers. My name is Naeem, and I am an engineer and maker myself,I am currently a Fab Steward at the South End Tech Center in Boston. I mostly make everything and anything I c... More About nemodude »

So recently my friend has gotten me a notebook with a cover that wasn't as appealing to me. Instead of being a jerk and getting a new book, I decided to create a new book cover for the book. After thinking about various book covers, I decided that the best course of action would be to make a book cover out of wood.

Materials needed:
- 1/8 inch sheet plywood
- old notebook (could buy new, but who wants to spend money?)
- glue (preferably craft glue)
- laser cutter
-computer (for the design and cutting)
-heat gun

Add Tip Ask Question Comment Download

Step 1: Design

First i gutted pages from the book cover. Then I hopped onto libre office draw (one of the many laser cutting softwares I'm comfortable with) and I drew living hinges.

Living hinges are little slits in wood or cardboard that allow it to bend without the use of adding actual hinges. I created different test hinges to see which best fit the book pages.

Add Tip Ask Question Comment Download

FIGURE 6.1–6.4 Naeem's Instructables post for making a wooden book cover with a living hinge.

Step 2: More Laser Cutting + Bending Cover

After I found the best hinge (not shown), I stacked them to the hight of the original book cover and added rectangles for the front and back of the cover. With the file made, I sent them to the laser cutter to cut.

After the cover was cut, I noticed that it naturally stayed in an open position. To get it to close naturally, I soaked the hinge in water for three minutes. Because I didn't want to bend the front and back cover as well, I applied wet towels over the hinge itself. Once three minutes passed, I bent the book to where it stayed closed and applied heat to it via heat gun so that it stayed in that position

Add Tip   Ask Question   Comment   Download

Step 3: Fitting and Glueing

Now all thats left is to glue the cover to the book. Being that this was the first ever book cover I made, I didn't know the best way to glue it. I do plan to learn a better method and put it on an updated instructable.

Add Tip   Ask Question   Comment   Download

FIGURE 6.1–6.4 (Continued)

his engagement in the social and intellectual practices in making. We briefly highlight two of these relationships: a) engaging selves—appealing to affective dimensions that support others' risk taking; and b) distributing expertise—acting as part of a knowledge ecosystem to which others also contribute.

After he made the actual object, he shared his experience with the Instructables.com community. As instructions go, the words and images he used to explain the project were not very detailed. There are other Instructables.com posts that include every single step of every process. However, Naeem's post reflected his design's stage of development at the time, not necessarily his final iteration. He explicitly stated in the post that he would add more detail later. Though his instructions were not finished, he opted to post anyway and he attracted the attention of thousands of viewers.

Making, through the culture of rapid prototyping, challenges notions of when an idea is ready to be published. By posting early in his process, Naeem could have an audience, gather ideas from comments—as he did—and work on future iterations. Composing the post and sharing it became part of his problem solving process for making a better book cover. He simultaneously posted to represent himself as a maker, seek recognition of his work, and to learn.

Engaging Selves

In our studies, we have found that a central aspect of making for makers is sharing what they know with their communities (Gravel, Tucker-Raymond, Kohberger, & Browne, 2018). Sharing is also central to Naeem's practice. Although there were book covers other people had made, and there were other examples of living hinges on Instructables.com, Naeem posted his project because no one had put the two together quite like he had. He had unique knowledge to share. Naeem had created new knowledge about engineering book covers. He shared his work as a way to inspire and invite others to participate and make. When asked in an interview why he chose to create the post as he did, he said,

> It was partly because I didn't want people to shy away from a cool project using a lot of words. That's a lot of text and that's good if you're going step by step, but I just want to get the idea out there of, "Hey, I made a wood covered book." I guess it's just more of the idea of not just all the steps . . . since it was a work in progress I left it at saying "I'm going to have more, I just haven't got to it yet."

Naeem was strategic about wanting to elicit an effective response to a "cool project." He wanted to expose his viewers to the idea, instead of possibly intimidating them with a detailed, text-heavy step-by-step description. His post was a way to share his new knowledge and invite others in the community to take a risk and make something as well.

We believe Naeem's role as a teacher at SETC contributed to the way he composed his Instructables.com how-to. Naeem's job in the Fab Lab at SETC was to engage anyone who walked in, to get them excited, as he was, about making things. It was an integral part of his making because making and teaching happened side-by-side in the Fab Lab. Through his post, he attempted to get others excited about his idea by "put[ting] it out there." Thus, Naeem's composition was not a demonstration of all that he knew about the subject, as is required so often in schools, but was socially oriented, sharing knowledge with an audience even though it might not be in final form. He posted to engage and inspire others in addition to getting feedback from them, an inspiration that in turn spurred his own making. His approach to composition also reflected the ethos of open access to knowledge, tools, and expertise that making communities share in general.

Distributing Expertise

Through his post, Naeem engaged in sharing in a way that implicitly acknowledged the distributed nature of knowledge and expertise in maker communities. Naeem's text did not exist as a standalone how-to. Other web pages on Instructables.com showed how to make other kinds of living hinges; Naeem showed how to make a living hinge for a book cover. Naeem's post existed as part of an ecosystem of how-tos, but also as part of a network of social relationships. It was a text among similar texts, some of which were complementary to his. For example, one reader who responded to his post pointed out that there was another maker who had a really good post about gluing wooden book covers to the pages. Even when sharing final or almost final products, products are open to feedback and there is always room for improvement or enhancement.

In making spaces, it is expected that knowledge is distributed among many people. Makers are interdependent because they need and expect others' expertise just as they know that they should contribute their own. Naeem implicitly trusted that his readers would leverage the distributed knowledge of the community and that they would find the information they needed about living hinges. His contribution helped to sustain a community that relies on exchanging knowledge and providing social support to achieve each member's goals.

Genre: Crowdfunding

Online crowdfunding has become a popular way for a variety of people to realize their dreams for risky projects without having to secure a bank loan. Often, hundreds of people contribute small amounts of money, sometimes for gifts or the product itself, to support a product or idea they believe in. Projects from participants in our study include raising money for writing a book about the jewelry of the Burning Man festival, creating movable magnetic comic strips for creating one's own comics, and manufacturing mini-PCR machines that can replicate DNA and test for disease without the use of a full biology lab. This last example we discuss in this chapter.

In 2017 there were about 200 crowdfunding platforms in the United States alone (www.fundly.com, 2018) and about 2,000 across the world (www.statistia.com, 2018). As of this writing, 15 million funders on Kickstarter, one of the largest crowdfunding sites, had given about $3.8 billion to support almost 150,000 products (www.kickstarter.com, 2018).

Tips from Fundly.com and kickstarter.com include doing research on your product area to be able to explain why yours is unique, and telling a compelling story that explains to potential backers who you are, what you want to make, the origin of the product, and why they should care. Crowdfunding hopefuls are highly encouraged to have multimedia, including a good project image and a compelling video in their campaigns. Fundly.com reports that those with videos raise 105% more than those without videos. They also encourage creators to be specific about where the money is going by including the budget and the production plan in their post. They should share updates regularly, offer rewards, and promote the campaign on other social media platforms by taking advantage of one's personal networks.

In short, the crowdfunding genre is a hybrid genre of personal narrative and persuasive essay with some exposition related to process. It resonates with what Skunk and Ecco, some of the artists at Artisan's Asylum told us: people do not just want to buy an art piece, they also want to buy a story. Seb, who interviewed with us about the product he made, the mini-PCR (Polymerase Chain Reaction) machine, said that after creating the crowdsourcing campaign, he

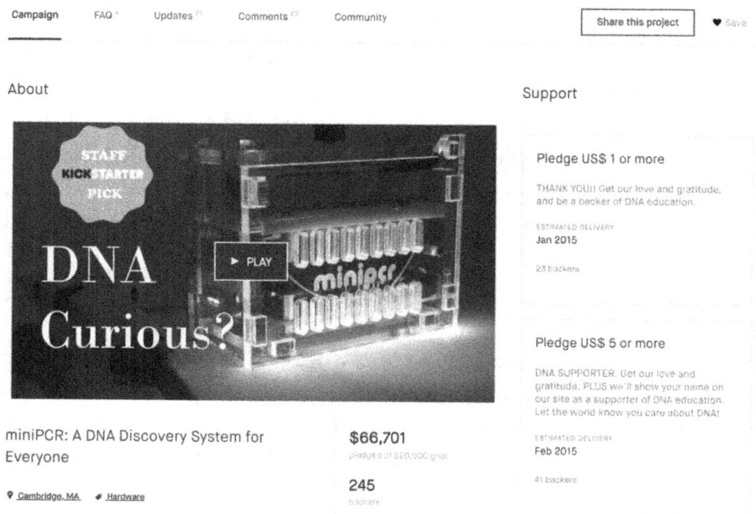

FIGURES 6.5 AND 6.6 Parts of the Mini PCR Kickstarter campaign page.

FIGURES 6.5 AND 6.6 (Continued)

realized it was not just a place to sell his product, but also a place to tell the story of his company and product,

> I have never seen Kickstarter as a way to communicate with your audience as honestly as you can. But it very quickly became that because it's a very unique medium. So, my previous view of Kickstarter is "it's a financial tool; it's a fundraising tool." My current view of Kickstarter [is that it] is a way to communicate.

Seb interviewed with us as an expert maker, although he was trained as a biologist, not a fabricator. He and his partner Z. created a miniPCR machine that copies DNA and allows anyone from students in a classroom to biologists in the field to do genetic analyses, such as look for bacteria in food or match DNA at a crime scene. Creating a crowdfunding campaign for him was not just a culminating activity, but one that helped him make sense of his own product:

> So communicating for me is a way of figuring out, "Is there a need and what [makes] defining that need better?" And that's something that I think I still struggle with. The closest I can say I came to feeling that way was when we wrote our kickstarter campaign and then we really put a lot of effort into we are about, you know on paper and text with images on video. It's very powerful.

For Seb, communicating his company's ideas through the kickstarter helped him and his business partner to refine their story and in doing so, to refine the need that they were addressing with their product. In creating their campaign, they went through a number of steps that helped them to make sense of what they were doing and to communicate that clearly to others.

> It was a process. I think we started with text on a page, and built a story line, you know, this is who we are, this is what we're doing, this is why we're doing it, what do you think, why you should help us, and then that started to take more physical shape. First we added a layer of images then we created some form of graphic design of concepts around DNA. And we layered that on top of the plain images. Then we decided to shoot a video. And feedback was huge. In the beginning we thought we were ready to put the thing live but then we sent it out to five people. And then we got a lot of [feedback] and then we sent it out to 10 more people and people helped us hone in and refine it and be clear.

Seb explained the importance of critique in creating a better story and the iterative nature of that story's creation. Creating the story through iteration and layering is much the same as creating a good product in the first place. It requires multiple cycles of research, design, and feedback. In preparing young people for sharing their work in public, it is important to help them anticipate how the public will respond to their work—both their product and their story about it. In his interview, Seb pointed out,

> When you put yourself out there. When you put your products [out there] it gets very personal and some people are going to be very supportive some people are going to be very indifferent, and some people are going to be very negative and aggressive. And we had already discovered the indifferent ones and the somewhat negative people. But we found the very aggressive ones through Kickstarter. People started to directly attack us. Some people got pissed off. They feel like you're doing something that they own and it's something that is so counter to everything that you think about. The maker movement and the open source world but I mean the world is full of personalities and you also have to be ready to deal with that when you're communicating.

As we noted, the crowdfunding campaign that Seb and his partner created was a hybrid genre that has roots in both persuasion and personal narrative. To set the tone, they used their first names in the post, implying an interpersonal familiarity. They also, through their video and text, asserted that there is a scientific community of MiniPCR using biotechnologists and educators that the funder can be a part of by supporting the work. For instance, part of their page stated, "Get MiniPCR for yourself but don't forget to give one to a school. Help us equip these public schools with DNA technology." In this statement, as in the rest of their post, Seb and his partner created a collaborative effort to "help get this product to market." Once their project was successfully funded, they included regular updates about their product, keeping the backers informed.

They also made sure to budget for the rewards they gave backers. It is important for creators to remember that they will also have to manufacture any rewards (such as product) that they give to backers and that those rewards cost time and money. At least one of our other participants indicated that he did not factor in how much the rewards would cost and he ended up breaking even on profit, but also spending an extra year and a half manufacturing and mailing all of the rewards, much to the dismay of his backers.

The MiniPCR creators also connected the importance of their product to individuals and then those individuals' relationships to the global community. The MiniPCR campaign placed DNA study in the context of people's ancestries, of disease, and in genetic engineering. In short, the campaign argued that by buying into the product, backers were becoming members of a worldwide scientific community committed to learning and creating a better world.

Finally, the MiniPCR team provided evidence for backers that their product was sound. In the video they shared that they have validated the product in different biotechnology labs. They included testimonials from practicing scientists and science educators. They presented to the audience their credentials as degreed scientists with publications and patents and they provided some of their high profile customers including NASA, MIT, and the Massachusetts Biotechnology Education Foundation (MassBioEd) (See Figure 6.6). In sum, creating a crowdfunding campaign is the process of figuring out one's story, telling that story, and convincing others that the story is worth buying into.

Genre: Blogging

Blogs are ways of sharing one's thoughts or, in making, what one has produced. They can also be an authentic way for learners to document what they are learning as they move through the process of making. The blog from the young people at the South End Tech Center required that they record their activities with the aim of inspiring or helping someone else with their own projects. It reflected youth's participation in a setting that valued learning, open source access to information, and a community that was connected across time.

The blog followed a set of conventions for most days. In the beginning, the team included a picture of themselves, a short description of their project and purpose, and the technologies they were going to use. Another component of the blog included a video of the group talking about their initial prototype. Much like the crowdfunding campaign, the blog was a way for the team to iteratively make sense of and refine their story. Making spaces are learning oriented. Blogs, and the other genres we discuss here, reflect that orientation. Online sharing genres in making contexts position both audience members and online composers as learners. Positioning sharing online as a learning activity rather than a final culmination of a project can encourage risk-taking for people doing the sharing because they are not expected to prove that they are already capable knowers and doers.

The blog also served the function of documenting what the team did each day. They posted videos of trials; in the morning, what goals they had for the day; and at the end of the day, what goals they reached or still need to work on. The team, WaterWorks, also included hyperlinks to resources that they used: other people's water pianos, code, and materials. Documenting in this way took on a distinctly social character. It was documenting not only for oneself but also for others. Not as an external requirement, but a necessity for telling one's story.

One of the expert makers that we interviewed, Helena, a mechanical engineering undergraduate and cosplay enthusiast, discussed a helmet she made with a visual equalizer on the visor that responded to music. In Cosplay—costume play—participants often dress as fantasy, manga, or science fiction characters. Helena wore her helmet to cosplay meetups, conventions, and on Halloween. As part of documenting her process, Helena shared her work online in a blog.

Helena had attempted to make the helmet through a type of fabrication process she had not done before, by creating a mold. She was not satisfied with the

Waterworks

Waterworks

Project

A musical water fountain that changes color when different Jumbo Piano keys are pressed.

It's Purpose

Its a way for children to be introduced to music interactively and develop a creative mindset using children's love of water in the hot summer as motivation to get them into music.

Project Technologies Utilized

1.Physical programing

2.Electronics

3.Digital Fabrication

4.Graphic Design

FIGURES 6.7 AND 6.8 Parts of the Waterworks team blog.

Proof of concept Design review

A panel of people who had more knowledge of the world technology came to give us feedback on our design.

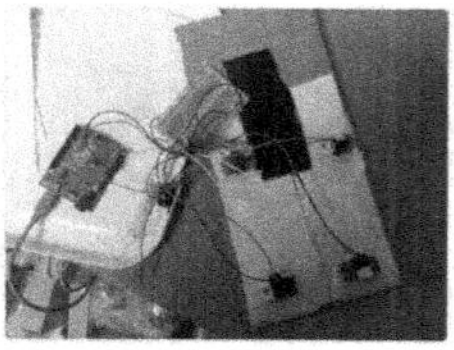

When my group tried showing off our prototype, we had some technical difficulties and it just wasn't working completely. But we managed to describe our prototype without the actual demonstration of it working.

The panel gave us really good feedback and suggested that we actually incorporate the water idea with the actual keys so the kids actually get a chance to play with the water.

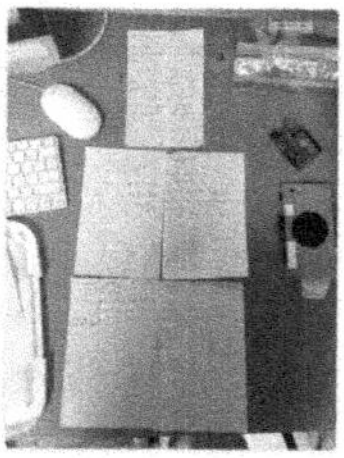

After the design review, we went back to the drawing board and rethought our water fountain idea to include water in the keys.

After a long group discussion we decided to include water in the keys for the kids to touch and by touching the water, the LED strip will change colors according to the notes.

We drew a rough blueprint of how we wanted the new piano to look. 

By the end of the day our group came up with a final blueprint and we wrote out our next steps to creating our project.

FIGURES 6.7 AND 6.8 (Continued)

blogs and instructions she found on the web and wanted to create a complete record of her process that would show the difficulties she had in completing each step.

> I made this mistake in the first couple of products that I ever made where I only took progress shots after I had finished a process like, "Look what this hat did." "Look what this process did." But it turns out that the final product doesn't really give a lot of insight as to how much work you put into it. And what all was necessary to do it. You look at that helmet, you can't imagine what the mold looks like, you can't imagine how long it took. You can't imagine what it looked like when it was a sad, plastic, white thing. And so, just taking pictures of the entire process has been incredibly useful.

Helena hoped that her documentation would be an educational resource for others who were attempting the same process. She explained her processes to others in case they wanted to replicate what she did. She wanted to help others learn from her missteps. It was also useful for her in reflecting on the techniques she used.

Helena's statement was representative of experienced makers in our study and reflective of what the youth in the WaterWorks team were doing with their blog. They all valued sharing what they made with other people. Their personal making was both for themselves and others, contributing to a community of people such as cosplayers or future SETC Youth Teachers. That stance also valued open sourcing. Open source access is a central value of the maker movement and stands in contrast to the proprietary nature of academic or market-based work in which intellectual capital and property resides with individuals. Information is freely shared on much of the web and many people share precisely because they believe everyone should have access to information. When one is learning to make the free flow of knowledge is indispensable.

Sharing Online: Compositional Genres in Making

A focus on the online genres of communication that makers use is simultaneously a focus on sharing genres' diverse modalities and purposes. Genres, such as crowdsourcing campaigns, offer real world opportunities for students in schools, and for makers out of school, to practice persuasion and storycrafting in multimodal contexts. Storytelling can help makers make sense of what they are doing.

Online genres in making are multimodal. They almost always employ images and words together. Sometimes those images are moving and sometimes they are static. Language can be written or spoken. The arrangement of images and words are integral to the message. People want to see what they are learning about and they want examples of how a process is undertaken. Multimodal compositions give creators opportunities to practice and develop facility with the affordances of different modalities. Through practice, makers learn what images and writing can convey best and how they can be arranged together to grab attention, encourage interaction with the consumer, and ultimately, teach. As Seb talked about, learning what conveys information best often comes from the online feedback and critique that makers receive when they share.

At the same time, shared online maker texts involve identity making in a way that is much more integral than in school-based writing. Online, makers create for a real audience that depends on them to be knowledgeable, understandable, and likable. Writing online implicates one's own identity and those of others, involves reciprocity (putting oneself out there), and provides space for learning, development, and belonging. In learning spaces, the production of online making genres can be used as part of authentic performance assessments that help to document what people learned. They can create an argument for their product, in the case of crowdfunding; provide a written record of activity, as in a blog; or create explanations for someone else to participate in the same set of practices as the learner had, as in how-tos.

Attention to the production of a shareable written text at the end of a project also encourages and gives purpose to documenting, one of the hardest literacy practices for makers to do naturally and one that many describe as having to force themselves to do. Makers can document through photographs and video and real time blogging. The Learn 2 Teach, Teach 2 Learn program has dedicated a person to documenting young people's activities and sharing them online. In formal settings, the teacher might act as a supporting documentarian of student projects. It gives the teacher evidence of learning as well as gives the learners opportunities to remember what they did and their decision points at different stages in the process. Integrated documentation also takes some of the burden off the learner to pay attention to so many moving parts at once. The Open Portfolio project at Maker Ed (https://makered.org/opp/overview/) is working to create a common framework for documenting, sharing, and assessing learning through portfolios.

Finally, sharing is a sensemaking activity not just a culminating activity, as pointed out by Seb earlier in the chapter. At the South End Tech Center, youth projects must undergo two early design reviews with panels of more experienced mentors. The first is after their initial ideation and a second is after they build the first prototype. Doing so helps them to figure out the audiences they want to reach, their constraints, and due to the reviewers' questions, helps them to think in more detailed ways just how their project will function. Sharing can become a regular part of the workflow before the final showcase or public engagement. The practice of crafting the story of their object and receiving feedback will help makers refine their stories and their projects so that when the final public engagement does occur, they will be ready.

The cases here suggest a set of considerations for designing social spaces to promote the development of digital literacies in making as a type of engineering practice: a) addressing or featuring problems generated by students; b) developing young people's networks of expertise; c) promoting the interconnectedness of texts; and, d) attending to social and affective dimensions of literacy engagement—including play and risk-positive presentation. These design considerations value and promote open access to information for communities to work together. In doing so, they can help youth develop agency as they seek information, share knowledge, and teach others. Situating sharing as sensemaking requires a community-oriented environment that treats individual learning as part of the greater, interconnected whole. We argue that these considerations can help to expand STEM literacy practices so that those in STEM learning settings consider social relationships as integral to learning. Digitally based literacies, from this perspective, can be simultaneously transformative for individuals, communities, and STEM spaces.

Making literacies are socially oriented, which stands in contrast to the ways that individual achievement is rewarded in schools. For instance, as a part of communities, many times people share online because they want to give back to the people who helped them or contribute to the sites where they learned. Learners' communities extend to their cultural communities, including their families, friends, and making spaces. Online texts encourage social and distributed problem solving. The interactive capability of the web allows those teaching and those learning to talk to each other across vast geographic and social distances. The wide-reaching character of the web often means that makers draw on multiple sources of information, integrating them into their own projects. Expertise is distributed across communities of interest. People are sharing into a world in which everyone is expected to know something and no one is expected to know everything. Texts and sources of information are hyperlinked to each other on sites and then dynamically linked to each other through users who move between texts and communities. Even the sharing texts that makers put out are not stand-alones. Cross-platform promotion of oneself and one's work is important to get the word out.

Guidelines for composing online making genres (how-tos, crowdsourcing campaign, and blogging) are included in the Appendix.

References

Barton, D., & Hamilton, M. (2012). *Local literacies: Reading and writing in one community.* New York, NY: Routledge.

Bhatia, A. (2018). Interdiscursive performance in digital professions: The case of YouTube tutorials. *Journal of Pragmatics, 124*, 106–120.

Gravel, B. E., Tucker-Raymond, E., Kohberger, K., & Browne, K. (2018). Navigating worlds of information: STEM literacy practices of experienced makers. *International Journal of Technology and Design Education, 28*(4), 921–938.

Heath, S. B. (1983). *Ways with words: Language, life and work in communities and classrooms*. Cambridge: Cambridge University Press.
NGSS Lead States. (2013). *Next generation science standards: For states, by states*. Washington, DC: The National Academies Press.
Swarts, J. (2012). New modes of help: Best practices for instructional video. *Technical Communication, 59*(3), 195–206.
Tucker-Raymond, E., Gravel, B., Wagh, A., & Wilson, N. (2016). Making it social: Considering the purpose of literacies to support participation in making and engineering. *Journal of Adolescent and Adult Literacy, 60*, 207–211.

Online Sources

https://blog.fundly.com/crowdfunding-statistics/
www.kickstarter.com/help/stats?ref=global-footer
www.statista.com/statistics/497227/number-of-crowdfunding-platforms-globally-by-region/
Autodesk screen shots reprinted courtesy of Instructables and Autodesk, Inc. and Naeem Wilson
Kickstarter screenshots reprinted courtesy of Sebastian Kraves and Kickstarter, Inc.
Parts of this chapter were reprinted courtesy of Wiley Publications.

7

TEACHER LEARNING IN MAKING SPACES

"By Design I Realize"

> **Focus Literacy in This Chapter**
>
> **Creating representational forms and traversing representational systems and materials:** *Producing symbolic representation of any aspect of the project as an aid to making (measurements, notes, drawings) or transitioning between different modalities such as talk, sketches, and formal designs.*

In this chapter we describe teacher learning through a case study of two mathematics teachers. The teachers were participants in a series of professional development workshops focused on STEM literacies in making that we organized at Nedlam's Workshop. Year Three of the STEMLiMS project focused on ways to integrate and study STEM literacies in making within a school context. Bringing making into schools requires collaboration with classroom teachers. Thus, we outlined a series of professional development opportunities for teachers at the school to support their learning around making, and to study how teachers came to understand the relationships between making and the role of STEM literacies in their disciplines. Following that fall workshop, we invited all the teachers to attend a semester-long teacher study group to think about integrating making into the classroom. This culminated with a two-day summer working session in which teachers built "kits" for other teachers in their school to support efforts for students to learn how to use the making space. Teachers in these sessions came from mathematics, physics, biology, and English. We focus on one learning experience between two mathematics teachers because it illustrates both how playing with representations and creating them across different modalities can lead to robust learning. We also focus on this case because the teachers bring up many of the same concerns for doing making-based activities with 30 students in a classroom that were raised by other teachers in our workshops and study groups.

Teacher Learning

The literacies in this chapter focus mostly on the ways in which teachers created and traversed representations as part of their own learning, both about their students as learners and also about themselves as learners and teachers. Understanding of teacher learning and professional development around student driven-construction and computational practices is still nascent. However, past scholarship has argued that for teachers to integrate disciplinary practices into their teaching they must experience some level of competency and agency with the practices themselves (Ball & Cohen, 1999). The same goes for technologically driven practices (Mishra & Koehler, 2006).

We argue that teacher learning about disciplinary making requires attention to the computational tools and infrastructures available to facilitate this work, as well as the ways teachers formulate, maintain, and constantly negotiate their own relationships to the disciplines and contemporary tools used therein. The effects of engaging teachers in disciplinary practices and thinking is clear in both science and mathematics education: teachers learn to develop professional vision in their own practice (Hill & Ball, 2009; Rosebery, Warren, & Tucker-Raymond, 2016), improve their teaching expertise, and deepen the learning of their students (Carpenter, Fennema, Peterson, Chiang, & Loef, 1989; Saxe, Gearhart, & Nasir, 2001). Finally, teacher inquiry into problems of their choice rooted in their classroom practice (i.e., teacher research) has also been shown to nurture teacher learning and agency (Ballenger & Rosebery, 2003; Zeichner, 2003).

Efforts to push making into the school curriculum require serious consideration of what Hawkins (1974) called the "I, Thou, and It." I, Thou, and It are the interactions between teachers, students, and curriculum. Where public schools are facing unrelenting pressures from standards and accountability directives, the intellectual, physical, and professional space for building disciplined inquiry into the curriculum is limited. We argue that teacher learning about making requires attention to the tools and infrastructures available to facilitate this work, as well as the ways teachers formulate, maintain, and constantly negotiate their own relationships to the disciplines and tools used therein and the people with whom they are working—their students.

Setting

This chapter describes a high school in Gateway Town, an industrial suburb of a major city in New England. The 165 full time faculty at the school served a student population of approximately 1,800. It included about 70% students of color, 43% students on free and reduced lunch, and 20% of students classified as Limited

English Proficient. With a few others in the region, the town served as a welcoming city to newcomer families in the United States. There was a significant immigrant population from every continent except Australia and Antarctica. At the time of our study, there were only a few classes in which students were making things. Computer science and auto shop were the most prominent. In the second semester, we would work with a physics teacher, Ashley Freewoman, to create an elective class called *Creative Design and Engineering* that would incorporate STEM literacies.

The making space, described in the introduction, is a 3,500 square foot former woodshop. Increased budget cuts and declining interest left the woodshop program dormant over the past few years and it served as a storage room where the occasional extracurricular club would come to do small, messy projects. In 2014, the district partnered with Tufts University, where Brian is a professor, to apply for and receive a National Science Foundation grant to create a new making space in the old woodshop. That new making space focused on "engineering inquiry" and forms of digital making. Over time, different tools were accumulated to support various forms of design and fabrication. The space opened first as an after-school program, facilitated by faculty and research assistants from Tufts, and then eventually the district assumed management of the space, placing Ashley as the new "makerspace" teacher. The push to explore the use of this space, and some of the practices of making that it engendered was felt from the beginning. Teachers and administrators were especially interested in how it applied to their normal classroom contexts. In the fall of 2016, teacher interest and support from the administration helped set the stage for the three different professional development opportunities that we helped create with teachers.

The first two-day workshop took place in November and December. It was intended as an introduction to acclimate teachers in the space, get them making things, and introduce the idea of STEM literacies in making. The second PD activity consisted of a semester long teacher study group in which we asked teachers to create and try out curriculum in their classrooms. Finally, the third PD activity was a two-day workshop in the summer that focused on teachers creating kits for teachers who wanted to a) integrate making into their classroom activities and b) explore the making space with their students. The events in this chapter are from the first workshop.

Participants

TABLE 7.1 Participants and disciplines in each of three learning opportunities at Nedlam's Workshop

Two-Day Workshop—Introduction to Nedlam's	*Teacher Study Group—Integrating Making into Curriculum*	*Two-Day Summer Workshop—Building Kits*
4 mathematics teachers	2 physics teachers	4 mathematics teachers
2 physics teachers	2 biology teachers	2 physics teachers
	1 mathematics teacher	2 biology teachers
	1 English teacher	1 English teachers

Rethinking Mathematics—Two-Day Workshop

This chapter describes and explores a conversation between two high school mathematics teachers, Kim and Georgia (pseudonyms). In their conversation, they are designing and creating 3D objects for the first time. The activity of designing 3D objects helped them think mathematically about the world and provided possibilities for helping their students think mathematically as well. Printing a 3D hexagonal prism with a cylindrical hole (see Figure 7.1) helped them to rethink a mathematics problem they assigned in their classes every year, namely, how to calculate the remaining volume of the hexagonal prism that was not cut out by the cylindrical hole. In the workshop, the teachers successfully made their first 3D printed object. In creating the physical model, they understood the mathematics problem in new ways. However, they did not think the activity would be useful for their students.

Mathematics is meant to address both theoretical and physical problems related to quantities, probabilities, patterns, spaces, structures, and changes in those phenomena. Likewise, theoretical, physical, and digital tools are used to solve problems and learn mathematics. In the younger grades, physical manipulatives such as Cuisenaire rods or pattern blocks are often used for learning. As students mature through the grades, they learn to use protractors, calculators,

FIGURE 7.1 Screenshot of the prism with the cylinder touching the sides. From the design program.

and eventually more powerful computers. However, as one progresses through the grades, the mathematical content becomes increasingly abstract. Those abstract ideas and formulas may become more tangible and concrete though making. Kim and Georgia were interested in finding out.

Kim and Georgia had almost 50 years of experience between them. Perhaps not relevant, but they were also both foreign born and raised their families in the United States. Kim immigrated from Korea and Georgia from Russia. At the time of the PD and for some years before, they both taught four sections a day, split between geometry and pre-calculus.

The first day of the workshop consisted of all participating teachers and project team members making a three-dimensional letter of the alphabet (such as an "H") out of paper. The first attempts to make a letter (all were between 4–12 inches tall) took about two hours. There were not any further instructions, but we provided graph paper, cardstock, and construction paper along with cutting and adhesive materials. We purposefully chose paper, scissors, x-acto knives, and tape as materials because of their familiarity. We also purposefully chose to leave the activity open ended where the goal was the same, but allowed everybody to approach it in their own way.

The second half of the workshop introduced digital tools (i.e., Inkscape and die-cutters) and asked teachers to continue thinking of ways to complexify their letter designs using simple circuitry or the new digital fabrication tools. At the end of the day, we convened a closing circle discussion, where we asked each member to write on a sticky note and remark on something they learned, something that surprised them, and a connection to the classroom. Kim shared how she learned "new project ideas—using a computer to design" and was surprised by how much time each task took to be done well. Kim said she found the workshop helpful because it reminded her that they could have students design objects and calculate area, volume, surface area, and perimeter as well as explore proportionality and transpositions of shapes, all through making 3D letters.

Georgia shared how their work building letters reminded her of an activity she did with students in her class in which they cut equal squares of varying sizes from the corners of pieces of paper to construct boxes of different sizes. The dimensions for each cut out represent a function that is related to the volume of each box. At the end of the day, Georgia wrote that she was surprised "by how many connections there were to her geometry class." She said she learned that hands-on activities such as making the 3D letter needed to be well prepared. She also thought students would learn more "using math concepts for real-life situations." We do not know if she thought what she was doing was "real life," but it was more than the abstract equations and paper and pencil problems she was currently providing her pupils.

Kim shared how she felt when they got started: "Actually I have no idea. Zero idea about how I can make a 3D whatever, so its <***> and I had a very good experience because I put myself in the students' situation." Kim went online to look for ideas for the letter she would make and eventually picked "H" because she thought it would be easiest. In making her letter she also thought about herself as a teacher. In a group discussion after making the letter, she said,

> And really the struggle process as I see it, if I tell the student "do all this," what limited instruction do I have to give to them in order to make the time more efficient in a forty-five minute class? Or maybe even ongoing. So what are the minimum steps I should [use to] guide them? For instance, tell them to explore the Internet and you have some ideas first and then design whatever you want to do according to the 3D model. Because I am teaching geometry, the 3D model is very very important and I think that this is a very good one because you can give them the proportion. Because I do it and the proportion pops up right away, maybe because I am a math teacher teaching geometry for so long. Because I will divide the proportion: okay this is this and what should be the proportion divided for the H, the two lengths and what will be the proportion of the cross bar. The whole idea, for example if I decide to draw a 3D model what will be the proportion. I think my students can get a sense. And then how to secure it, how to flip it. The whole process I really like it. From the students' point of view, I like it.

Kim saw value in the activity of creating a three-dimensional model, relating it to key topics like proportion and ratio. Yet, the newness of the activity in her teaching gave Kim pause. She was not sure how long it would take or what some of the road bumps might be for students. However, by putting herself "in the students' situation" she could imagine the kind of work they would need to do to be successful. She also saw the world mathematically in her work, "Proportion pops up right away, maybe because I am a math teacher teaching geometry for so long," while seeming to intimate that her students may not see this immediately. The letter making activity most certainly provided representations for thinking mathematically for the teachers. Kim and Georgia thought about them through proportion, transformations, area, surface area, volume, perimeter, and the ways in which functions worked in real life situations.

On the second day of the fall workshop series, two weeks later, we offered three mini-workshops where teachers could learn about woodworking, 3D printing, or using a computer-controlled die cutter that cut shapes out of paper. Four of the teachers chose woodworking and went off to build stools that could be assembled without metal hardware. Two mathematics teachers, Georgia and Kim, buoyed by their experience making three-dimensional shapes two weeks earlier, wanted to learn how to use the 3D printer. The challenge was to come up with something that they could design and print using the design program tinkercad®. One of our research team members, Aditi, helped them through tutorials learning the new representational system of the software.

When introduced to the program, Kim and Georgia first wanted to make two dimensional shapes, like triangles, that students could play with, learn proofs about, and eventually make their own. For their first 3D printed object, Kim proposed that they print a hexagonal prism that would act as a container, and a cover to go along with it. In the beginning, the teachers approached using the 3D design software using the traditional forms they knew, two-dimensional configurations. Using this new tool, with new affordances and new representational systems, required thinking about the representations of their disciplinary pedagogy in new ways. In this case, it required considering moving from two dimensions to three.

Kim and Georgia wanted to make a hexagonal box with a lid. This required making a small hollow depression in the top of the hexagonal prism where a similarly-shaped bump on the bottom side of the lid could rest, holding it in place. As Kim was learning how to create that hollow depression, she came to a realization: the hollow shape she was imagining, a cylinder, could actually be the same height as the hexagonal prism, it did not have to be a small slice that only created a depression. This realization of the ways shapes could be combined and subtracted led her to consider something slightly more complex. She wanted to make a hexagonal prism with a cylindrical hole hollowed out of the middle. It was a model of a classic problem that they gave their students each year.

KIM: Now I have an idea. We have an actual lesson. We have a hexagonal object. But we have a hollow cylinder (gesturing the cylinder inside the prism).

GEORGIA: It asks what will be the difference in volume.

KIM: The difference in the volume when the hole is taken out, [and all that is left are] the little edges. What is the volume left? So now the same idea but now I'm not going to insert the same hexagon only more, I'm going to insert the cylinder. That is exactly what we are teaching for unit 8.

ADITI: Do you want to do that?

KIM: Yeah, because I play around and out pops the idea. We can actually do that. And we can actually (looking at hands), Okay (looks at Georgia), calculate the volume for the cylinder. The model of the hole of the cylinder.

GEORGIA: (Nods)

Kim and Georgia came up with an idea for an object to print that would fit a problem they typically asked students to solve. Up until this point, they had only worked on the problem with students using mathematical notation on paper—equations and perhaps some sketches. But by learning the software and its different affordances, by being in a space with low consequences (Georgia's stated goal was simply to produce "something") they were able to "play around" and make connections to what they were teaching in class. They thought designing and printing the object would help their students think mathematically about the real world.

It took a little time to learn to get the solid and hollow shapes to align in the program, but once they did, the pair started manipulating the cylinder to make it as big as possible. When it touched the edges of the hexagon exactly, they stopped. They considered it a while. Then, Aditi explained that the 3D printer would be extruding plastic layer by layer and when the round hollow cylinder touched the edges of the hexagon, there were no sides, because the hole had taken their place. The edges of the object would be erased or they would be so thin that they would not be able to keep their shape and stand up.

GEORGIA: It will be too thin.

ADITI: Yes, too thin.

KIM: So smaller, okay (touches keyboard) But now we run into the other problem because the actual calculation of the volume should be actually touching here. [What we do in class] is not realistic. It's not realistic I know that. Actually, now I see it. Actually, when we talk about it, this one, we have to be more realistic now. Because usually we calculate it [as it] touches the hexagonal side. We always teach [it this way]. We forgot that if you touch the side, actually, the side there, [is] cut off. It becomes disconnected.

GEORGIA: But probably if you leave more [of the hexagonal] and take it out it won't touch.

KIM: This is the reality now though. I realize we never [thought] about that before this year. This is the diameter of the cylinder so that will also be the radius of the hexagonal. But now we forgot we have to leave space in order to hold the volume (smiles and laughs quietly). By design, I realize. All these years I didn't realize that.

In all their years of teaching they had never considered how the math problem they were figuring out in the classroom would apply in a real-life situation. It was like doing mathematics in a vacuum. Conceptually ideal, in that the circumference of the circle matched the radius of the hexagon, but "not realistic" as Kim said. The design created a tension between providing students with ideal situations to simplify the mathematics and providing "ideal" problems for learning and making the mathematics authentically applicable to the real world.

Designing for fabrication, moving between different representational forms of the problem, encouraged Kim and Georgia to reconsider the mathematics they taught and the models they used. Designing a 3D printed object encouraged them to reimagine quantities, shapes, and structures with a variety of representations. For example, one of the features of tinkercad is that it treats a shape as a "solid" or as a "hole," and both are objects according to the program. They are a part of a menu of choices that can be picked up, dragged, and resized. A machine (e.g., a person with a hand mill, CNC router, and, in this case, printer) needs to know the dimensions of an object in order to make it, either by adding or subtracting material to a substance. If one wants to hollow out a shape, then, the outer dimensions and the dimensions of the hollow must both be accounted for. There needs to be at least two sets of representations, even if just measurements written in centimeters or inches. Yet, Kim and Georgia, and presumably their students, had been considering the hexagonal prism with the cylindrical hole as one shape, ready-made. The design program helped to turn the "hole" into an object. That is, an object of consideration in its own right. It was not the means to the ends (the volume of the remaining hexagon), rather an equal partner with a volume of its own.

In their work together as 3D designers, Georgia, Kim, and we, positioned the tools as resources for helping them think about how to present problems to students, problems where the current modes of representation were unsuccessful in helping students grapple with the mathematics. Georgia remarked that she was struggling trying to support pre-calculus students in understanding logarithms and that all she was thinking about in this moment, standing with tinkercad and the 3D printer, was how to better help them understand it. While it is unclear how she was seeing the role of 3D printing in understanding logarithms, we do believe Georgia was commenting on the different modalities of representation that she was learning and how that might expand her thinking about new possibilities for representations in the curriculum.

Yet, despite that curiosity, Georgia was skeptical of the activity's value in the high school classroom, "How is this applicable if we have 50 students who have never done it?" Kim thought that since they had made the object, they might now be able to draw on their own experience to discuss how to design such an object with students. However, both agreed it was time consuming and would take disproportionate time away from the rest of the curriculum. They also worried about keeping the students on task if there were too many frustrating moments, fearful students might give up and not learn the math.

GEORGIA: I don't see how that is going to fly in a 30 student class.

ADITI: If students are working in pairs and they know something about the basics of using the software and designing you'd want them to bump into challenges like this. They come up as points for discussion and then facilitate.

GEORGIA: (nods)

KIM: Well, not in this design though because it takes too much time, you are right. (points at Georgia). 30 students, 34 so how do you actually get them to the same place. It is very very difficult. We can actually uh* . . .

GEORGIA: We have lots of other things so I don't know. It's a challenge.

KIM: It is it is.

GEORGIA: It's time consuming.

KIM: Time consuming. Also, control the chaos (looks at Aditi) of you know—there is just two of us wondering how are we going to make [our students do what we did]? If we have 34 students divided by two or even four we have several groups. How can we? . . . As a teacher you don't want them to just run wild just explore on your own. You don't. You always want

> to [make sure they are doing] some instructional thing. I could use the projector to make them do what I teach them [and] at least they can follow. But I myself have to be really familiar first, otherwise I will run into trouble. "How do I do this, I don't know!?" (laughs)

Kim and Georgia raised several challenges: a) ensuring students all learned the same thing, or made it "to the same place"; b) fitting making 3D models in among the many other demands on time; and c) too many students and/or groups of students to watch over. A central theme that tied these three challenges together for the teachers was their discomfort with not being experts on the tool. Georgia stated that there were, "many details to take into consideration" with the software program that one did not know about until one understood what the program could do. Another was that they needed to be in control of their students. Such statements indicated a traditional teacher-led stance toward classroom learning. Taken together, we infer that the teachers thought that to be successful they would need a tight curricular focus and a solitary voice of knowledge embedded in the teacher to keep things from getting chaotic.

They also acknowledged the complexity of teaching. There were many moving parts, and for them to do it right, they needed time to plan lessons and write down the program functions so students could progress through the software at an even pace. Kim and Georgia thought it would be a good idea to help them become familiar with the tool by continuing to play with it over the course of the school year, and then they would have enough time to plan for the following year.

Kim and Georgia were new to 3D design and printing. Although they each remarked that, in prior years, they had students make physical objects in class, neither had done so in a long time, classifying it as a possible generational problem. They said people their age did not want to break the tools and so they were careful when learning new skills. They were both still apprehensive about using the technology in the classroom. They want to completely know the tool before using it, which is understandable given the way the distribution of knowledge and identity works in schools. The design immediately made Kim reconsider a standard problem she had given her students, and it challenged her view of the curriculum. However, using new, digital representational and fabrication systems was not something that she felt she could do because her students all had to learn the same content. It seems the values of making—distributed expertise and networked knowledge—were at odds with school.

Renegotiating Relationships

Bringing together their disciplines, pedagogies, and students, Georgia and Kim's three-dimensional letters made from paper, as well as the hollow hexagonal prism in its different forms, represented mathematical models that they and their students could think with to understand relationships between volume, shape, proportion, area, surface area, and perimeter. Representations are important in mathematics, central even. Certain topics are presented so tightly coupled with conventional representations that we often forget to examine those connections—between ideas and the forms they take in the world. Many times those representations are limited to the page.

Pushing their thinking about representations to three-dimensional objects helped the teachers decouple the representations they had always used with students from the mathematical understandings. The representations that they used were not limited to the page. Rather, they included talk, hand-drawn diagrams, the design software, and the printed plastic hexagonal prism with a "6" on the side [Figure 7.2].

They also had different kinds of familiarity with representations, which constrained what they were able to with them. For instance, they knew to sketch on graph paper, draw distances and angles, and make notations. However, they had to learn a new representational system in the design software. And when they first started, they continued to think in two dimensions, in slices not shapes. To some extent, their learning a new software was supported by their facility with the familiar representations of their sketches and talk. However, they also needed to make the jump from two dimensions to three. They had to figure out how to put a hollow cylinder inside a hexagonal prism. This kind of translation, trying to capture and express an idea across different forms, is foundational to developing mathematical understandings (Goldin & Shteingold, 2001).

Thinking about this in the context of making, it was playing with the representations and creating objects in the real world that led Kim and Georgia

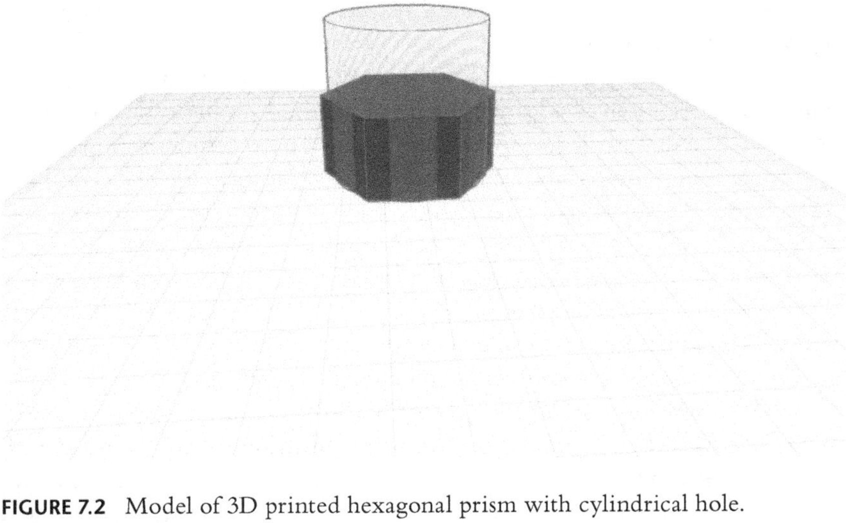

FIGURE 7.2 Model of 3D printed hexagonal prism with cylindrical hole.

to change their ideas about the mathematics they presented to their students. Play is as important for teachers as it is for students. Bruner (1972) noted two important aspects of play: As a "means of minimizing the consequences of one's actions, and of learning, therefore in a less risky situation," and as "an excellent opportunity to try combinations of behavior that would, under functional pressure, never be tried" (p. 693). Play creates space to take risks, explore boundaries and structures, and to put one's own practices into conversation with objects and with other individuals (Vygotsky, 1978). Considering play within disciplinary contexts, Hawkin's (1965) cited "messing about" as an important mode in science, where children are encouraged to explore materials and conjecture about their behaviors; playful interactions with materials creates opportunities for learners to "bump up against the world" (Bransford & Schwartz, 1999, p. 93), where they may attend to moments when the world does not behave as they expect.

Goals for our workshop included wanting teachers to be excited about the possibilities that the making space and tools could offer them and their students. However, even though the experience challenged and changed Kim and Georgia's ideas about the mathematics they taught, their time spent with the tool was minimal and there may not have been enough opportunities to directly rethink relationships to students. For instance, there were no students present. Therefore, Kim and Georgia did not envision that they would be using 3D printing and design in their classroom. They expressed very real constraints of time and classroom management.

However, we argue that part of the realization of those constraints is because Kim and Georgia were continuing to think of the ways in which 3D printing would fit into their existing math classes as if knowledge and information would continue to flow in the same ways, in ways that maintained a limited distribution of expertise and where the teacher is the knower. The characteristic value of making spaces including "distributed expertise" and "learning oriented" suggests a repositioning of roles in these spaces, encouraging teachers to understand themselves as learners alongside students, as both contributing to the production of knowledge, even if this way of constructing relationships to knowledge is not immediately obvious to teachers. We offer that teachers need to reconsider at least three dimensions for learning through making to be successful in schools: a) their disciplines, b) the tools and representations that they use to teach those disciplines, and c) their students. As such, we argue that successful teacher learning and making integration into STEM courses in schools results from professional learning opportunities for teachers that encourage them to explore new relationships to those three dimensions.

In this illustrative case, we believe that our professional development workshop allowed teachers to consider disciplinary questions as they learned new tools and representations. However, the workshop experiences did not do enough to consider how teachers could change relationships to students and knowledge. In Hawkins's "I, Thou, it" formulation, we missed the "Thou." In part, a focus on play and exploration did not allow for a critical rethinking of who could know and how information could be disseminated in classrooms. However, experiences of co-learning, co-constructed learning spaces with students may help encourage the distribution of expertise among community members. We recognize the challenge of involving students in setting learning goals and activities in the face of mandated and tested curricular content. However, we believe that there is enough space in everyday classrooms for teachers and students to explore making with multimodal representations that answers deep disciplinary questions.

References

Ball, D. L., & Cohen, D. (1999). Developing practice, developing practitioners: Toward a practice-based theory of professional education. In L. Darling-Hammond & G. Sykes (Eds.), *Teaching as the learning profession: Handbook of policy and practice* (pp. 3–34). San Francisco, CA: Jossey-Bass Publishers.Ballenger, C., & Rosebery, A. (2003). What counts as teacher research? Continuing the conversation. *Teachers College Record, 105*(2), 297–314.

Bransford, J. D., & Schwartz, D. L. (1999). Chapter 3: Rethinking transfer: A simple proposal with multiple implications. *Review of Research in Education, 24*(1), 61–100.

Bruner, J. S. (1972). Nature and uses of immaturity. *American Psychologist, 27*(8), 687.

Carpenter, T. P., Fennema, E., Peterson, P. L., Chiang, C. P., & Loef, M. (1989). Using knowledge of children's mathematics thinking in classroom teaching: An experimental study. *American Educational Research Journal, 26*(4), 499–531.

Goldin, G., & Shteingold, N. (2001). Systems of representations and the development of mathematical concepts. In A. A. Cuoco & F. R. Curcio (Eds.), *The roles of representation in school mathematics* (pp. 1–23). Reston, VA: NCTM.

Hawkins, D. (1965). Messing about in science. *Science and Children*, 5–9.

Hawkins, D. (1974). I, thou, and it. In *The informed vision: Essays on learning and human nature* (pp. 48–62). New York, NY: Agathon.

Hill, H. & Ball, D. L., (2009). The curious—and crucial—case of mathematical knowledge for teaching. *Phi Delta Kappan, 91*(2), 68–71.

Mishra, P., & Koehler, M. J. (2006). Technological pedagogical content knowledge: A framework for teacher knowledge. *Teachers College Record, 108*(6), 1017–1054.

Rosebery, A. S., Warren, B., & Tucker-Raymond, E. (2016). Developing interpretive power in science teaching. *Journal of Research in Science Teaching, 53*(10), 1571–1600.

Saxe, G. B., Gearhart, M., & Nasir, N. I. S. (2001). Enhancing students' understanding of mathematics: A study of three contrasting approaches to professional support. *Journal of Mathematics Teacher Education, 4*(1), 55–79.

Vygotsky, L. S. (1978). *Mind in society: The development of higher mental process.* Cambridge, MA: Harvard University Press.

Zeichner, K. M. (2003). Teacher research as professional development for P—12 educators in the USA [1]. *Educational Action Research, 11*(2), 301–326.

8

IMPLICATIONS FOR PRACTICE AND RESEARCH

As the title of the book suggests, our work, and this study, are intended to present compelling cases with implications for learning, teaching, and research. Making spaces are learning spaces, whether there are adults learning with other adults, youth learning by making projects for their community with adult mentorship, or students learning new skills and techniques together in design and engineering tasks. A goal of this volume was to advance theoretical positions related to the use of representations among communities of makers in making spaces. We mean these implications to be useful for a wide range of spaces, from dedicated 40,000 square foot making spaces to pop-up spaces inside classrooms.

In this concluding chapter, we describe opportunities and tensions for educators, teachers, researchers, and other people who design learning environments that are interested in promoting STEM literacies in informal, school-based, and hybrid spaces. We present a set of design principles distilled from our observations across four making spaces, give examples of how those principles might be put into action, and conclude with recommendations for future research.

Design Principles for Making Spaces That Promote STEM Literacy Practices

Our design principals center literacies to begin bridging learning in informal spaces with learning in more formal spaces. New forms of multimodal and multidisciplinary activity, cultivated in making spaces, can shift how educators think about learner participation in STEM practices. These design principles include the following: Authentic Communication, Representational Fluidity, Meshworking, Community and Reciprocity, and Distributed Expertise.

Authentic Communication

The principle of authentic communication is related to the literacy practices of *communication for help and feedback, sharing,* and *sharing for audiences.* By authentic, we mean people's communication for real and practical purposes in service of the making. The geographer Doreen Massey (2005) argued that space is the product of interrelations. There are many interrelations. Relationships between people is one type, and communication is inherent to building those relationships. That is, spaces are produced by interactions between individuals. Interactions are in turn shaped by the spaces they construct. As a design principle, this means healthy making spaces should be organized to encourage and foster authentic forms of communication.

In the making spaces we observed, interrupting was an acceptable practice. Yet, in schools, people are made to wait their turn, to work silently and alone, to sit in their seats. In the making spaces, work was public in a way that attracted curiosity and discussion. Our observations suggest that opportunities for learners to move around and talk to each other should be made available. At the same time, scaffolds that support asking for and giving feedback can be integrated into learning activities so that when students do ask for or give feedback, they have some resources from which to draw. Moreover, time not working directly on a project is not necessarily time off task. It can include talking to other people, socializing, sparking ideas, helping solve problems, hanging out, and building relationships.

Spaces that promote authentic communication provide opportunities for experienced and new makers alike to explore their identities as artists, engineers, makers, and STEM doers. Structuring communication in this way might be challenging, particularly in schools where some of these practices seem antithetical to accountability and performance narratives that focus on individual performance and external validation. We propose the use of sharing genres for making as one means for structuring how people communicate around the work they are doing in the making space.

Sharing Genres for Making

Chapter 6 discusses the role of online sharing in making. Oftentimes, makers come to understand their work by telling others what it is they are doing. Including multiple opportunities for storytelling allows makers to reflect on their work and forces them to articulate it for others. Whether it is through sharing, documenting, or communicating online, we noticed that makers produced recognizable kinds of texts, or genres, for different purposes in their work. Crowdfunding, how-to guides, blogging, and social media genres are important and authentic to making, because they are the forms that makers populate to express purposes, identities, ideologies, values, and particular approaches to audiences they are trying to reach, either to share a product they made, or knowledge about a topic.

Certain genres have been constructed to meet the needs of makers, whether to raise funds, share techniques, or seek help. In that sense, there is a process of creating different genres that poses opportunities for learners to engage literacies in STEM learning. Existing genres can structure how learners think about producing texts in making, offering guidance on the purposes of these forms, and opportunities to try out their ways of expressing and sharing what they learn while making. As a principle, this means exploring the established set of genres (e.g., how-tos, crowdfunding, blogs) as part of the end product of making activities in addition to constructing the artifacts themselves.

We developed protocols for youth makers that asked them to generate texts of these particular genres: Instructables for a project or a process, YouTube "how-to" guides for skills and techniques, Kickstarter campaigns to raise funds and awareness for a project, and the regular use of blogs and other online media that can be applied in different contexts (e.g., blogging about literature, politics, or anything, really). We found that the production of these texts led to youth clarifying and refining their own senses of the skills or ideas in question. They used them to make sense of what they were doing. In Nedlam's Workshop, two students made a first-person "how-to" for putting a screw into a piece of wood. As they began filming, they realized they had not fully understood how to use the chuck on the drill to loosen and tighten a drill bit or driver bit. They then had to think step by step about what they were trying to do, and they had to practice. Making a tutorial, which amounts to populating the genre, also required they watch other YouTube tutorials as models of short "how-to" videos. Thinking about how others produced tutorials also became a moment to explore the idea through the production of this text.

We found other instances where creating an Instructables post, or writing a blog post, surfaced the need to clarify bits of confusion for makers. As we think about the role of literacies in STEM, in particular the production of texts as means of sense making and expressing ideas in particular forms, then adherence to the authentic genres of the activity—in this case, the genres of making—is a way to encourage the use of texts that support and enhance the makers' own goals, values, and ideologies as they develop. Storytelling is sensemaking. This positioning of textual production as purely authentic and in service of the work runs in tension with some conversations around the documentation of making in schools.

Tensions with Documentation

As making activities move from the fringes of conversations about learning in schools to the center of those discussions, the role of documentation in making quickly rises to the surface. How can documentation be made authentic? How do educators know what a person learned? Documenting is a literacy practice that we observed in making spaces. Although the adults said it was the most difficult part of their practice to keep up with, we recorded many examples of documentation at Artisan's Asylum and at LMNPO. Decisions were recorded, solutions to particular coding or circuitry problems were written out and diagramed, and an attention to keeping track of important information by jotting notes or creating lists was a common practice among the experienced makers. As we moved into youth spaces, SETC and Nedlam's Workshop, we saw fewer examples of either spontaneous or deliberate documentation. We did see facilitators and adults taking pictures and sharing them with youth. We saw some notes drawn (e.g., the "map" in Chapter 5 that WaterWorks produced could be considered a form of documentation), some links recorded in blogs, and comments left on code in online repositories. However, largely speaking, documenting was less prevalent among youth makers than among the experienced adults. And yet, conversations about making in schools continually come back to documentation.

It seems that documentation is attractive in conversations about making in schools for at least two key reasons. First, documentation is a means for collecting moments of work, processes, and thinking for the purposes of evaluation and assessment. It is hard to assess what youth have learned, in the ways school frames both assessment and "learning," looking at projects like WaterWorks. Documentation becomes a practice that yields the bits and pieces plugged into the assessment engines (e.g., rubrics, portfolios, etc.). Second, school has long been organized around producing texts, and there is a tight coupling between how we conceptualize learning in school and the production of texts. As though figuring out how to get small solar panels to charge a cell phone is not learning until one can write that down somewhere in complete sentences.

We are sympathetic to the call for documentation in school-based making, there certainly is a space for this, but we argue that documentation should not be motivated solely for the purposes of assessment, or some other outside evaluation of the work. While we saw some benefit to the encouragements of sketching in design journals in Nedlam's Workshop, we caution against the requirement of these kinds of documentation. Similarly, perhaps it is not a requirement for youth take pictures of their work, but rather it could be the work of the adult educator to collect moments of making on camera. Instead, documentation should be an authentic practice in the service of providing material for the sharing genres. They can be used to help structure the collection of artifacts and records. If it is expected that a how-to video, or an Instructables post, will be produced at the end of some making project, then that reframes how learners and educators think about what to record and capture. Sharing genres are part of the structure of the activities from the beginning and they can motivate the documentation for reasons authentic to the ways makers capture and share their work.

Representational Fluidity

In Chapter 5 we introduced the idea of representational fluidity. Representational fluidity is the practice of *traversing representational forms*, conventional and emergent, in ways that support making. Another term that has been used, representational fluency, would suggest the maker has facility with various conventional forms, and a capacity for moving among them by constructing relations between them (e.g., Nathan, Stephens, Masarik, Alibali, & Koedinger, 2002). But this does not help us explain how *new* forms are created, explored, or understood in the moment or the ways people might move between invented and conventional forms of representation. In representational fluidity, representations are objects to think with (Vygotsky, 1978). A representational fluidity approach centers representations as powerful and necessary tools. The metaphor of fluidity suggests that representations are in motion, as people move through projects, they pick and choose different representational forms to fit their needs. At the same time, representations do not become the domain of any one person, but represent the collective production of knowledge, highlighting the multimodal and multidisciplinary values of making spaces. Fluidity then becomes a characteristic of a group of people or of a space, not of an individual.

Representational fluidity blurs conventions and inventions and foregrounds the ways representations aide thinking, becoming partners in the iterative refinement of ideas and designs in making. Although we recognize the power of conventional forms of representation (e.g., Cartesian graphs, C++ syntax), fluidity allows for unconventional and invented forms as well. Fluidity is thus important because of the power that is located in the ability to manipulate representations in STEM fields, stratifying participation based on familiarity of conventions and formalized systems. Access to canonical STEM discourses is limited when these discourses only include historically accepted forms of the disciplines and do not leave room for hybridity, informality, or play.

We would argue that making spaces, where representational fluidity is practiced, promoted, and celebrated, expand the points of entry to these privileged forms that often serve as gatekeepers to STEM participation. A space that encourages inventions—maps of Arduino pins, or Sharpies drawn on 3D printed prototypes—promotes multiple ways of knowing and producing knowledge. Acknowledging these power differentials in how representations are positioned when we build and operate making spaces can help in the effort to create spaces that are more equitable, and that promote the participation of people from groups marginalized in STEM fields.

To build spaces, facilitation practices, or approaches to making that embody representational fluidity means paying special attention to the role of sketching, invention, and traversals in the activities in which participants engage. We explored this with simple interventions at SETC and Nedlam's Workshop. One such approach to emphasizing the constant and central role of sketching, like we found at Artisan's Asylum, was to place clipboards around the making space. Each was mounted with clean white paper and a black felt-tipped pen. It sounds specific and perhaps simple, but the contrast of the thick, firm black lines on white paper is appealing. And we found that mediating each and every conversation with the youth makers by asking them to show, draw, or sketch their ideas on the loose pages encouraged sketching as a practice that youth eventually began to do on their own, without prompting. In school contexts, "sketching" or producing drawings of any kind outside of art means producing the *right* kind of drawing, like a free body diagram in Physics, or a graph in math. It is not unusual for students to assume that there are right, proper, or accurate ways of sketching, which is an idea that can be challenged in the context of making spaces. We also provided design journals or notebooks that makers could use in whatever way they pleased, but the primary goal was to bring sketching into the fold as a regular, creative practice. Finding ways to encourage sketching as an informal and personal practice begins to dismantle notions that there are right and wrong ways to sketch. Given the emphasis on problem solving in STEM, a focus on sketching, and sketching across materials, supports how makers see and understand problems; it supports their developing fluidity with representations of many forms as they become tools to think with during design and making activities.

Meshworking

In Chapters 3 and 4, we emphasized the role of relationships between individuals in making spaces, the ways they communicate, and the ways information flows through these webs of relations. The literacies we explored were related to *communication for help and feedback* and *identifying, organizing, and integrating information.* As we say in Chapter 3, we initially conceptualized this as a network, with nodes representing individuals and lines symbolizing the relationships. The metaphor of networks focuses on structure, or how we understand the constitution of individuals as the infrastructure through which knowledge and information is passed and constructed. In networks, people are the nodes. Information flows between them. However, this network view leaves some questions with respect to how ideas, people, and processes emerge and evolve in making spaces, and it also limits our understanding of how particular practices develop over time.

We can imagine networks as a static representation of spaces yet we have argued spaces are inherently dynamic. Thus, we introduced the notion of the "meshwork" as an alternative to networks. Ingold (2011) pushes us to question the idea of nodes in a network, because that amounts to an inversion, or a collapsing of attributes, experiences, and contributions into the individual rather than situating these features of life in the relations between individuals, the lines of the meshwork. Thinking about work in lines shifts how we think about how individuals produce making spaces. It is not only their presence in the space that contributes to the values, ideologies, customs, and activities of the space, but also their lines of life that bring them into and out of that space. These lines of life flow across and through the making space.

Dynamic meshworks can shift how we think about what knowledge is valued and how it is collectively produced and shared in the space. Part of the work of making spaces is to seek ways to make these existing meshworks visible for makers, particularly youth, and to encourage the continued development, growth, and expansion of these meshworks. We acknowledge that within formal contexts, like school, one's view of the meshwork will be conditioned by institutional frames. In other words, divisions like "school" and "home" are felt by students in schools, communicated through so many messages from the school, and therefore we see meshworking as a possible way to begin troubling, and dissolving, some of the distinctions between the different spaces of learners' lives (Leander, 2002).

We piloted a design specifically intended to support youth in coming to see their existing meshworks, while also beginning to grow them (we provide a protocol for this in Appendix C). Through a process of "meshwork mapping," we asked youth participants at Nedlam's Workshop to begin by thinking about who they go to for help with questions or information inside the school context. We began there as a means of acknowledging the ways being at the school would frame youth's ideas about seeking help, asking questions, and gathering information. Participants placed their names in the middle of a piece of paper. They then wrote the names of individuals or places that they go to for help in school. The students included things like "parents," "teachers," and "friends." Or they parsed space into "School," "Life," "Random Stuff." As they progressively filled out "maps," talking with peers about kinds of information they sought, kinds of questions they asked, or similar kinds of sources they used, details emerged. Kyle wrote the names of some of his closest teachers, his guidance counselor, the person who handles job placements for students, people he goes to about "random stuff," and one person who helps him with issues related to "law" (see Figure 8.1). He also listed "Building Stuff," presumably because this took place in the making space during a summer program in which he was building shelves. Kyle, and some other students, began to talk about people in their lives who they interact with around this idea of building stuff, or making things. Alex, a 10th grade student, talked at length about his father, writing "He knows a lot of things," in reference to his father on his meshwork map. That list of "things" was elaborated over time to include electrical work, framing for houses, flooring, and other kinds of contracting or house maintenance skills.

As a design principle, we advocate for the careful attention to how relationships are formed, and how makers' dynamic web of interrelations, or meshwork, reflects shared knowledge, expertise, and access to different kinds of

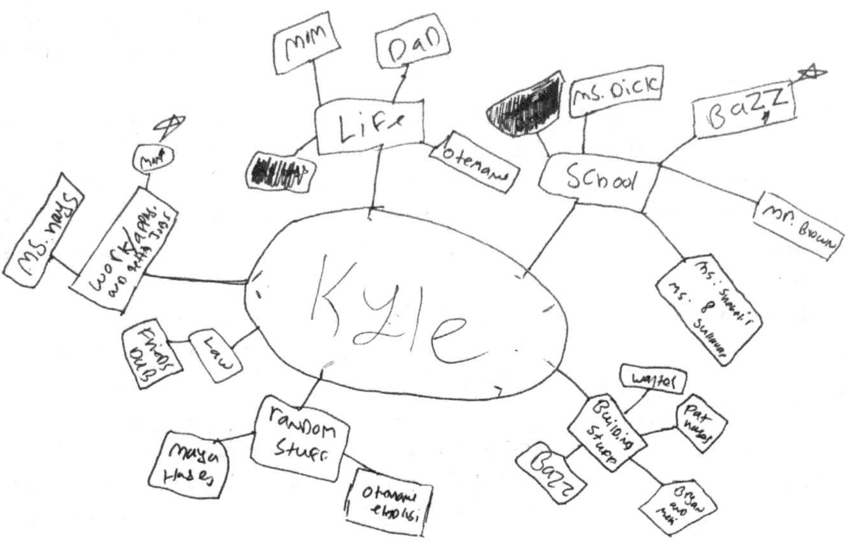

FIGURE 8.1 Photograph of meshworking map.

information within and, most importantly, outside of the making space. For example, encouraging projects that solve community needs or address problems in youth's lives position the work of making in a way where one's involvement in local meshworks becomes a resource for designing and solving problems that have consequences for those very same communities.

Community and Reciprocity

One of the most striking elements of Artisan's, SETC, and LMNPO was the community orientation to making, sharing, communicating, and helping. Individuals relied on their communities for support and, in turn, supported their communities where they could. We elaborate on this idea by describing some of the characteristic values in making spaces that contribute to community-oriented perspectives.

Open Sourcing is one such characteristic value of making spaces. We define it as the free sharing of ideas, tools, and resources, including community and local resources that have specific values among members, similar to "local food" movements (Hatch, 2014). Open sourcing helps to democratize production by putting it in the hands of individuals through no/low-cost shared access to expensive or high tech tools for everyday people (e.g., Buechley, 2006; Gershenfeld, 2007). The commitment to this principle by the makers we observed spanned from the ways they shared ideas and products online (e.g., GitHub), to the ways they offered help and feedback to others in their community (see Chapter 3), to the ways they choose the specific tools to use. For example, at

Artisan's Asylum, the would-be executive director and friends built their own laser cutter and offered the plans and guides for use by anyone, rather than purchase one from a distributor.

Open source structures allow access and shared production. Individuals can find what they need but they can also contribute to shared knowledge. Communities use, produce, and maintain the information that members need. Open sourcing is a principle that prescribes how information is situated within a making space. We advocate for the open and free sharing of information with all spaces, in both formal and informal settings. Posting training and resource guides on open Google Docs, for instance, allowing unrestricted access to tools and materials, and intentionally considering the role of information support (e.g., cascading mentorship, or "help desk" hours) are ways of committing to this principle in the formulation and continued operation of a making space.

Some designs we relate specifically to democratization of information in making. With respect to identifying, organizing, and integration, our designs focused on exploring the notion of "open-sourcing" as a foil to information navigation practices within schools, where particular messages about copyright, plagiarism, cheating, and the proprietary nature of information dominate. We wondered how we might be able to use making as a means to push on how the school constructed the role of information for students. One of the confounding effects of this exclusive and proprietary view of information is that youth have limited experiences searching for information online in ways that yield helpful results. We described in Chapter 4 how socializing is a useful practice for navigating some of the challenges of information seeking and evaluation, but we wanted to explore designs that could support how youth tap the "open source" nature of the maker world to learn more widely useful practices for finding information.

One activity we explored was "co-searching." Co-searching involves a simple arrangement where a more experienced maker sits next to an emerging maker, both with computers in front of them (or phones or someway to connect to the Web), looking for information together. Having agreed upon the general topic, or question, each maker announces what they are using for a search term. This is one example of the intervention. We found that over time, supporting youth as they searched for information online was a way to communicate the value of open-sourcing, while also supporting developing IOI practices.

Spaces where sites like YouTube are restricted (every school district we have ever worked in) should reconsider censorship and ways of accessing information, to communicate a different message about how knowledge and information is positioned in the community. Furthermore, and we take this up in more depth later, the principle of open-sourcing means that the communities that construct these spaces must be oriented toward collective knowledge. One's contributions are made to enable the community to improve. Knowledge is the collective "property" of the community, and information is the medium through which interactions and relationships in making are forged and maintained. This is in stark contrast to the individualistic, proprietary version of knowledge and information for which schools, and the economic market, advocate. For them, knowledge is to be owned. For making spaces, knowledge is to be shared.

Distributed Expertise

Making spaces are contexts where different forms of work, knowledge, and skill can interact in novel and unusual ways. When jewelers and mechanical engineers work side by side, the intersections of their knowledge and practice become fertile ground for developing new ways of working and making. Making spaces often share elements of carpentry and machining workshops, artist's ateliers, and engineering and science labs. We observed this at Artisan's Asylum, where the community paid special attention to the collective knowledge and expertise of the membership. They have an electronics studio next to a jewelry shop and a CNC shop next to the woodshop. The distribution of expertise within multi-use making spaces places different historical forms of production in proximity to each other. This contributes to hybridity, where new ways of working and learning emerge, including new literacies that can bridge conventional representations with emerging and idiosyncratic ones. This hybridity nurtures fertile intersections of expertise, tool uses, and discourses to produce innovative practices and transformative approaches to making, learning, and being.

Makers explore unique arrangements and combinations of different ways of working and develop identities and social relations at and across these intersections. A tacit goal in forming a community in this way, as a collection of tools and people representing a diverse range of domains and interests, is that the overall range of expertise will grow and expand; the community will know more than any one individual could. That complex, broad-reaching expanse of know-how contributes to the conditions for distributed expertise. An array of possibilities exists within the meshwork of the making space, consisting of contributions of knowledge and practice to the overall collective resources of the community.

Different people know different things, and they support each other as they explore new frontiers in their making, be it new processes for manipulating new materials or new tools used for tasks previously unexplored (e.g., laser cutting forms for jewelry making). However, a collection of knowledge in the space does not mean it is distributed. Rather, spaces should endeavor to embody a participatory commitment and culture (Jenkins, Ito, & boyd, 2016). In participatory cultures, as opposed to consumerist cultures, the expectation is that individuals bring something of value to the community. People's participation contributes to the collection of knowledge and expertise.

We observed distributed expertise in action when we explored the ways information flows through the meshworks (see Chapter 3 and Chapter 4). This does not mean distributed expertise is at odds with individual learning. In fact, it's central to the characteristic value of the spaces being learner-oriented. Halverson, Kallio, Hackett, and Halverson (2016) frame it this way: "Participatory cultures are organized to coordinate collaborative inquiry toward solving unknown questions, distributing knowledge across the community, then toward specifying the next set of questions on the horizon" (p. 4). Making spaces cultivate a distributed expertise similar to participatory culture, where making and learning are the core activities, and collective inquiry emerges in how individuals support each other in their endeavors. In addition, although not necessary, the community can also begin to scope and determine new questions they wish to take on. As a principle, distributed expertise repositions the knowledge of the space as collective and participatory. Based on the principle of community and reciprocity, members are assumed to offer expertise and support as well as receive expertise and support when they need it.

Wall of Expertise

In making spaces it is helpful for people to know each other's names, interests, and areas of strength. When they run into problems, they can find the right person to help. New ideas can serendipitously emerge from different practices operating side-by-side. We have experimented with what we call a "wall of expertise" adapted from a practice we observed at the Olin College of Engineering. On the wall, each member of the making space is listed. Next to their names are descriptions of what tools they know how to use, what techniques they are good at, or what content they might know. Members can also include their level of expertise or openness to helping someone think through a task. People do not have to be actual experts to begin putting their skills on the wall of expertise. These walls serve at least three purposes: a) they allow other members of the space to find the right person to whom they can ask a question or get advice from, b) they give learners an opportunity to announce their developing skills and interests, and c) they give educators and facilitators a sense of what else makers might want to learn or where they might need to develop skills more deeply. At Olin, the wall of expertise is used to keep track of who has been trained to independently operate the fabrication equipment (e.g., laser cutter), and this acknowledgment of expertise requires taking a test. In our spaces, we have asked learners and educators to put up their own skills by writing on sticky notes and putting it next to their name on large chart paper. Either approach, or a hybrid, could be used. Whatever is done, distributed expertise depends on three supports: a) hybridity of the space through different tools and activities, b) the intentional development of skills and knowledge, and c) establishing methods for intentional connections between people.

Summary

We have presented five basic design principles that we argue support learning STEM literacies in making spaces. We have also given some practical ways in which each of the principles can be manifested in the design of making spaces. We believe these principles treat literacy as relational, as existing between and among people, representations, tools, and goals. Considering literacies from a relational perspective is understudied. However, there are also tensions in basing learning and teaching on relational perspectives in some spaces, such as schools where individual achievement is valued (Kuipers, 2011). We hold no assumptions that a relational perspective will be adopted wholesale by more formal institutions of learning. It will be up to educators, and future researchers, to figure out how these tensions will play out and be addressed. Next, we present implications for what this perspective might have on literacy studies, studies of making, and STEM learning research.

Implications for Research

In the beginning of the book we positioned STEM Literacies as multiliterate social practices. Others have also conceptualized literacy and STEM learning through a lens of social practices (Barton & Hamilton, 2000; Lewis, 2001; Roth, McRobbie, Lucas, & Boutonné, 1997). In this implications section we have further refined this positioning to account for literacies from a relational perspective (e.g., Burnett & Merchant, 2018; Hawkins, 1974; Ingold, 2011). As we argued previously, a relational perspective helps to conceptualize knowledge and information as the domain of the social group, tools, and materials that are present. Knowledge and information are created with the emergent production of the space, and they are not the property of individuals or the products of their skills or aptitudes (Lefebvre, 1991).

In Chapter 1, we also discussed multiliterate social practices as the unit of analysis between particular time-bound events and larger social ideologies, discourses, and material relations of power. However, if literacies are fluid, and people are moving through time and space, coming together dynamically, as our data suggest, then we need a new way of conceptualizing literacy.

> Just as context is fluid, hybrid, or even collapsed as Marwick and boyd (2011) claim, so literacy events are porous and permeable and may lack the patterned predictability of Heath's original conception (Burnett & Merchant, 2014). Moreover, the idea of patterned, rule-bound literacy events sits uncomfortably with notions of liveliness, affect, fluidity, and emergence.
>
> *(Burnett & Merchant, 2018, p. 3)*

Burnett and Merchant (2018) argue that literacy events are not bound by time and space. Our data suggest as much. Research directions might then focus on how literacy practices travel in and out of making spaces, where those other spaces are, and what kinds of values and ideologies cross those permeable boundaries. What kinds of analyses can researchers do that account for stretches of time, past, present, and future? How do we account for the emergence of activities, literacies, and invented or idiosyncratic representations? What kinds of designs might support representational fluidity, and how can those designs encourage the interrogation, sensemaking, and the use of different representational forms?

In this volume we built on the idea of meshworking to understand the ways in which people moved through time and space, came into contact with each other, communicated, and shared "entangled lines of life, growth, and movement" (Ingold, 2011, p. 63). Future research can build out this conceptualization to better understand how expertise is distributed and shared as well as how it emerges in practice and contributes to the formation of activities and the use of representations. If meshworks are composed of the tangled lines along which people act—and along which the lives of spaces, tools, and materials run their course—then we need ways of understanding and representing how those lines become entangled, what their relationships might be, and how they might support STEM literacy practices. Some literacy scholars have used Delueze and Guattari's (1987) rhizomatic analyses in studying spaces (e.g., Leander & Rowe, 2006; Leander & Sheehy, 2004). Rhizomatic analyses account for dynamic, emergent, and multiple trajectories of participation in and through spaces. Future research might focus on how rhizomatic analyses account for lines of becoming, entanglements, and emergent practices in making spaces and STEM literacy practices.

From a relational perspective, questions about the roles identity and affect play in learning with STEM literacies while making are worthwhile to explore. How do we account for affect, excitement, and frustration when people come together to engage in literacy practices? And, how do we account for identity in literacy practices? We have explored maker identity as it relates to sharing online (Tucker-Raymond, Gravel, Wagh, & Wilson, 2016), identifying, organizing, and integrating information (Gravel, Tucker-Raymond, Kohberger, & Browne, 2018), and lifelong learning (Tucker-Raymond, Gravel, & Pierce, in press), but how are maker identities implicated, produced, and renegotiated across practices in different spaces? How are identities both collective and individual? What kinds of implications might those productions and renegotiations have for the ways in which people participate in literacies?

A relational perspective also allows researchers to focus on the production of power in making spaces. Power is not something of fixed amount or pre-determined in relationships (Foucault, 1980). It can be fluid. Studies are needed that focus on issues of power in STEM literacies and in making spaces, that connect to the national and international contexts of marginalization and exclusion that permeate making, schooling, and STEM careers alike (Buechley,

2013; Calabrese Barton & Tan, 2018; Norris, 2014; Vossoughi & Bevan, 2014). Questions might include understanding how official texts, rules, and policies of making spaces and the institutions in which they reside contribute to the kinds of relationships that are built and the ways in which access to communication and information is facilitated. Likewise, understanding the ways in which making and STEM literacies contribute to and reproduce the marginalizing and inequitable economic formations of neoliberal markets and an innovation economy that benefits the few will be important questions if educators are to purse equity goals. Finally, along these lines, we are encouraged by the work of scholars who are pursuing questions of critical STEM agency among makers (Calabrese & Tan, 2018). More work needs to be done to connect processes of agency and structure in STEM learning spaces (Varelas, Tucker-Raymond, & Richards, 2015).

Chapter 7 highlighted how teachers began the process of renegotiating their relationships to tools and the curriculum through the exploration of their own making and its connections to their home disciplines. As we think about making in schools, and the role of facilitation in informal spaces, we have begun designing and researching professional learning environments that support teachers in renegotiating three components: curriculum, tools, and students. Questions we want to answer include the following: How might teachers engage in playful making to begin thinking about these relationships? What kinds of relational structures to each other, to students, and to professional learning can best support shifting their thinking? And, what kinds of epistemological shifts may be important for integrating making and STEM literacy practices into their classrooms?

Clearly there is much work to be done. We have only begun to think about what STEM literacy practices across making spaces might look like and how we might understand and support them. It is our intention to continue this work because of the value we see in such spaces and the opportunities for learning that can emerge in them. We also continue these lines of inquiry because of the opportunity for misuse, misappropriation, and co-optation that can happen when technology is centered in learning spaces. An expansive view of the relational and social production of STEM literacies can help address inequities in making and in STEM learning. We end with a quote from Susan Klimczak who is the education organizer at the South End Technology Center. She, too, points to the importance of social, reciprocal, and community-building stances in learning while making,

> It's not about an individual just learning things. One of the things we ask of everyone that comes in here is, "What can you teach someone else?" So there's this process of we're building knowledge together, we're developing relationships, we're learning how to work together. Those are all literal literacies that are very important.

References

Barton, D., & Hamilton, M. (2000). Literacy practices. In D. Barton, M. Hamilton, & R. Ivanič (Eds.), *Situated literacies: Reading and writing in context* (pp. 7–15). London: Psychology Press.

Bordieu, P. (2003). *Language and symbolic power.* Cambridge, MA: Harvard University Press.

Buechley, L. (2006, October). A construction kit for electronic textiles. In *2006 10th IEEE international symposium on wearable computers* (pp. 83–90). IEEE.

Buechley, L. (2013). *Thinking about making.* In Keynote presented at the Annual FabLearn Conference, Palo Alto, CA.

Burnett, B., & Merchant, G. (2018). Literacy-as-event: Accounting for relationality in literacy research. *Discourse: Studies in the Cultural Politics of Education.* doi:10.1080/01596306.2018.1460318

Calabrese Barton, A., & Tan, E. (2018). *STEM-rich maker learning: Designing for equity with youth of color.* New York, NY: Teachers College Press.

Deleuze, G., & Guattari, F. (1987). *A thousand plateaus: Capitalism and schizophrenia.* Minneapolis, MN: Minnesota University Press.

Foucault, M. (1980). *Power/knowledge: Selected interviews and other writings 1972–1977* (C. Gordon, Ed.). New York, NY: Pantheon.

Gershenfeld, N. (2007). *Fab: The coming revolution on your desktop—from personal computers to personal fabrication.* New York, NY: Basic Books.

Gravel, B. E., Tucker-Raymond, E., Kohberger, K., & Browne, K. (2018). Navigating worlds of information: STEM literacy practices of experienced makers. *International Journal of Technology and Design Education, 28*(4), 921–938.

Halverson, R., Kallio, J., Hackett, S., & Halverson, E. (2016). *Participatory culture as a model for how new media technologies can change public schools.* WCER Working Paper No. 2016–7. Wisconsin Center for Education Research.

Hatch, M. (2014). *The maker movement manifesto: Rules for innovation in the new world of crafters, hackers, and tinkerers.* New York, NY: McGraw Hill.

Hawkins, D. (1974). I, thou, and it. In *The informed vision: Essays on learning and human nature* (pp. 48–62). New York, NY: Agathon.

Ingold, T. (2011). *Being alive: Essays on movement, knowledge and description.* London: Routledge.

Jenkins, H., Ito, M., & boyd, d. (2016). *Participatory culture in a networked era.* Malden, MA: Polity Press.

Kuipers, J. (2011). Unmentionable others: Representing participation frameworks in school science. *Anthropological Quarterly, 84*(1), 87–99. George Washington University Institute for Ethnographic Research. Retrieved October 4, 2018 from Project MUSE database.

Leander, K. M. (2002). Polycontextual construction zones: Mapping the expansion of schooled space and identity. *Mind, Culture, and Activity, 9*(3), 211–237.

Leander, K. M., & Rowe, D. W. (2006). Mapping literacy spaces in motion: A rhizomatic analysis of a classroom literacy performance. *Reading Research Quarterly, 41*(4), 428–460.

Leander, K. M., & Sheehy, M. (2004). *Spatializing literacy research and practice* (Vol. 15). New York, NY: Peter Lang.

Lefebvre, H. (1991). *The production of space* (D. Nicholson-Smith, trans.). Oxford: Blackwell.

Lewis, C. (2001). *Literary practices as social acts: Power, status, and cultural norms in the classroom.* New York, NY: Routledge.

Massey, D. (2005). *For space.* London: Sage Publications.

Nathan, M. J., Stephens, A. C., Masarik, D. K., Alibali, M. W., & Koedinger, K. R. (2002). Representational fluency in middle school: A classroom study. In *Proceedings of the twenty-fourth annual meeting of the North American chapter of the International Group for the Psychology of Mathematics Education* (Vol. 1, pp. 462–472). Columbus, OH: ERIC Clearinghouse for Science, Mathematics and Environmental Education.

Norris, A. (2014). Make-her-spaces as hybrid places: Designing and resisting self constructions in urban classrooms. *Equity & Excellence in Education, 47*(1), 63–77.

Roth, W. M., McRobbie, C. J., Lucas, K. B., & Boutonné, S. (1997). Why may students fail to learn from demonstrations? A social practice perspective on learning in physics. *Journal of Research in Science Teaching, 34*(5), 509–533.

Tucker-Raymond, E., Gravel, B., & Pierce, E. (In Press). 80/20: Making identities in making spaces. In J. Gammel, S. Molutsky, & A. Rutstein-Riley (Eds.), *I am what I become: Constructing an identity as a lifelong learner.* Charlotte, NC: Information Age Publishing.

Tucker-Raymond, E., Gravel, B., Wagh, A., & Wilson, N. (2016). Making it social: Considering the purpose of literacies to support participation in making and engineering. *Journal of Adolescent and Adult Literacy, 60*, 207–211.

Varelas, M., Tucker-Raymond, E., & Richards, K. (2015). A structure-agency perspective on young children's engagement in school science: Carlos' performance and narrative. *Journal of Research in Science Teaching, 52*(4), 516–529.

Vossoughi, S., & Bevan, B. (2014). Making and tinkering: A review of the literature. *National Research Council Committee on Out of School Time STEM*, 1–55.

Vygotsky, L. S. (1978). *Mind in society: The development of higher psychological processes.* Cambridge, MA: Harvard University Press.

APPENDIX A

Methodology

This book draws from a three-year project called Investigating STEM Literacies in Making Spaces. It included two years of using ethnographic techniques, namely observations and interviews, to study four making spaces. The last year was spent designing and studying teacher professional development opportunities and a high school class at Nedlam's Workshop. Our techniques for data collection remained fairly constant with a heavy reliance on fieldnotes and interviews. However, in the second and third years, where we studied smaller spaces in which we could guarantee almost everyone had consented to the project, we employed more video and audio recordings. We used detailed fieldnotes as a guide to identify episodes for much closer analysis of talk and action in the spaces that we could examine in the recordings. When we started the project, there was an emerging body of work on making, but little research focused on making spaces specifically (Sheridan, Halverson, Litts, Brahms, Jacobs-Priebe, & Owens, 2014). Therefore, we wanted to create case studies of the spaces and the literacy practices enacted within each space.

Before we entered the making spaces, we wanted to both refine our list of literacy practices and get a sense of the breadth of practices we might encounter. We interviewed 14 experienced makers (five women and nine men). In each object-based interview (see Appendix D), we asked general questions about the makers' backgrounds as well as more detailed questions about an object that they made, or were in the process of making, that they felt represented their skill set and outlook as a maker. We also asked them about the literacy practices these makers may have used (e.g., if you needed to learn something to do the project, how did you find out what you needed to know?). While all of the people we interviewed were experienced at making one thing or another, we sometimes talked about an object they had made for the first time. Results from this study were published elsewhere and showed that literacy practices were distributed across the life of a project (Tucker-Raymond, Gravel, Kohberger, & Browne, 2017; Gravel, Tucker-Raymond, Kohberger, & Browne, 2018).

Data Collection in Making Spaces

When entering a making space, one can quickly be overwhelmed. People are often on computers, loud tools, talking, moving, shouting, and playing music. At all times, our data collection was focused on identifying and understanding literacy practices as they related to STEM topics. This focus, along with keying in on only some participants, helped us to wrangle the amazing amount of activity in each of the spaces. We employed four primary forms of data collection in this study: interviews, observations with fieldnotes, recordings, and artifact collection.

Interviews

We used interviews in each phase of the work, with participants in each of the making spaces, as a means for closely examining the ways individuals understood their making processes and the role of representations in their work. Two categories of interviews were developed. The first focused on objects and representations. Participants were asked to talk about their practices and processes anchored by the story of a single object they had made. We used this protocol with adults and youth. The second category of interviews focused on making in particular contexts and representations. For these interviews, questions were about how the individuals worked with others, what materials and tools they used, how they sought and offered help and feedback, and other aspects of what it means to make in a social space. These interviews often focused on particular projects as well. Some participants were interviewed individually, and others were interviewed in groups. All told, we conducted 40 interviews with adult makers and 22 interviews with youth makers.

Observations with Fieldnotes

Ethnographic observations were a primary data collection method for this study. Our methods built on two core perspectives on ethnography. The first was Emerson, Fretz, and Shaw's (2011) idea of ethnographic participation. We acknowledged our roles as researchers in the spaces we observed. We interacted with participants and talked with them. We asked them to clarify what they were doing, or to speak aloud their process. We understood our fieldnotes to be subjective accounts of making in the different contexts we observed. We took care to record quotes, to describe setting, and to construct thick description (Geertz, 1973) in our work. Our process involved the production of "jottings" in the field, followed by the writing of more complete and detailed fieldnotes that were entered into the data corpus.

The second perspective on observations and writing fieldnotes built from Heath and Street's (2008) work focused on language and multimodal literacies. Particular attention to representations was a consistent thread in our approach; this is a study about literacies, and representations were our primary mode of exploring those practices. Heath and Street call particular attention to the social aspect of language, noting that humans creating "symbolic structure for one another" (p. 3) follows patterns at times, and diverges into idiosyncrasies at others. Our observations focused on how individuals engaged in making, with representations, but they also focused on the interactions and interrelations we observed between interlocutors in the different making spaces. This attention to social dynamics, dialog, and cooperative work allowed us to better understand literacies as social practices. It grounded our observations and how the fieldnotes were produced.

Recordings

Making spaces are loud, messy, often frenzied, and generally difficult places to collect the kind of video data that has become so important to research in the learning sciences. And yet, without these records of participants at work, we have limited opportunities to micro-genetically describe the nature of interactions in making spaces (Jordan & Henderson, 1995). Video recordings are useful data in developing case studies, in examining particular interactions, and in helping to understand contextual features of particular moments we observed. Fieldnotes constituted our primary mode for collecting data on people making. However, we collected video recordings in different contexts to explore particular lines of inquiry. For example, we positioned cameras on youth working together on making projects to document dialog around objects and representations (e.g., looking at code on a computer screen together). We asked participants to wear head-mounted cameras to capture work from their perspective. We recorded whole-group discussions in the *Creative Design and Engineering* class, and presentations of work through design reviews or in final project showcases. We also recorded teachers engaging in discussions and playful making as part of professional development sessions.

Artifacts

Making is about the production of artifacts, digital or physical. As such, what people make is an important part of the story, and important data to collect. However, people also want to keep what they make, and the best record of these artifacts that we could generate were pictures of their creations. However, we also made a point to scan written documents (e.g., loose paper and pages from notebooks) and to download blog posts and other online documentation that participants produced, and we asked some makers to send us images of objects they had made in the past. These artifacts became part of the multimodal record of what our participants produced. It is quite difficult to account for the total corpus of artifact data. However, we were able to collect artifacts in the following forms: photographs, downloaded web pages, scans of sketches and notebook pages, message board correspondence, emails, and objects produced as part of interventions, like "meshwork maps" (see Chapter 8). These artifacts feature prominently throughout the book, as components of our analysis and as illustrations of the multimodality of making.

Data Collected in Each Making Space

TABLE BM1.1 Data collection in each space

Data Type	*Y1: AA*	*Y1: LMNPO*	*Y2: SETC*	*Y3: Teacher PD @ Nedlam's*	*Y3: Creative Design & Engineering*	*Totals*
Observations & Fieldnotes	96 hrs	102 hrs	98 hrs	49 hrs	35 hrs	380 hrs
Interviews	19	7	22	0	0	48
Recordings	5 hrs	3 hrs	40 hrs	44 hrs	70 hrs	162 hrs
Artifacts	Yes	No	Yes	Yes	Yes	N/A

Artisan's Asylum

Observations at the Asylum totaled approximately 96 hours from January to June. We interviewed 19 participants (interviewed and observed), including 4 staff members (volunteer president and paid directors of: membership services, facilities, and education and outreach) and 15 community members who represent a range of expertise including jewelry makers, sculptors, visual artists, mechanical engineers, electrical engineers, welders, digital fabricators, woodworkers, roboticists, tinkerers, and hackers. Two of the community members in our study are also board members. An initial list of participants who were long-time, day-time, full-time members was given to us by the site liaison to the project. We then used referential sampling to find more participants. Observations included the following: members individually making in their studio spaces or in shared shops; makers collaborating with others; people making objects for themselves; formal classes in digital fabrication, small scale metalwork, leatherworking, and woodworking; and organized informal hangouts in electronics and robotics, events that are open to the public such as city-sponsored events like artists' open studios. We also observed new orientation sessions as well as town hall meetings.

Recordings were minimal. We asked three people to record themselves with a head-mounted camera as they worked, two electrical engineers and an artist/engineer. We recorded two conversations of group collaboration. Artifact collection was opportunistic rather than systematic. We collected some designs from participants if they were relevant to the interview conversation or observation. These primarily included CAD designs, sketches, notebook pages, and photographs of work. We also collected data from personal and professional websites, including the Asylum website.

LMNPO

We observed at LMNPO approximately 102 hours from January to May and we interviewed seven of ten employees. Observations included design, fabrication, and tinkering processes as participants made a variety of objects including, among others, a large "touch easel" screen for young people with autism, laser cut boxes, interactive displays for a room at a museum, a small injection molder, various machined parts, a specialized coffee cup for single pour coffee, a day's travel case for diabetic medicine and supplies, PCB boards for electronics products, a kinetic sculpture for a university making space, and an electronic lovebird wristband that, when tapped by one partner, would be felt by another.

Video and audio recordings were kept to a minimum because LMNPO asked us not to record anything that was proprietary or in violation of a non-disclosure agreement they may have had with clients. We did record two design conversations and some work on the three dimensional CNC mill. We were also able to take a number of photographs of people at work. Again, due to the proprietary and secretive nature of the work we did not collect artifacts.

SETC

Observations at SETC totaled approximately 98 hours from November to August. From November to June we observed one day a week either on Thursday open drop-in studio time or during Saturday workshops for prospective Youth Teachers. The bulk of observation time occurred during the summer when youth were present four days a week. Observations included design and fabrication activities, team meetings of College Mentorz and Youth Teachers, teaching activities, and what the education organizer called "groupiness" activities in which youth would get to know each other. We conducted a total of 22 interviews, including two with the program's education organizer and one with the center's director. We had five object-oriented interviews, including two with College Mentorz and three Youth Teachers. We interviewed three College Mentorz and eight Youth Teachers about their teaching roles. We also held group interviews with four of the summer project groups about the objects they were fabricating and their group processes. We recorded over 40 hours of video and audio data during our observations. These included standalone cameras and head-mounted cameras that we asked youth to wear while making (see Chapter 4). We collected numerous photographs and copies of data and representations.

Nedlam's Teacher PD

We conducted three separate professional development opportunities for teachers in Nedlam's Workshop. All instances included observations and fieldnotes, audio and video recordings, and artifact collection.

- Two-day introductory workshop, five hours each day, for which we collected two days of fieldnotes (10 hours) and 11 hours of video and audio recordings, including standalone cameras and head-mounted cameras worn by two of the teachers. We did not conduct any interviews. Artifacts, and photographs of artifacts, such as teacher sketches, were collected.
- Semester-long teacher study group that met every other week. We audio recorded about 19 hours of group conversation and took fieldnotes for 13 meetings (about 19 hours). We did not conduct any interviews or collect any artifacts.
- Five-day workshop in the summer to develop classroom kits to introduce people to making. We took fieldnotes all five days (20 hrs) and recorded about 14 hours of video and audio. We did not collect artifacts or conduct interviews.

Nedlam's *Creative Design and Engineering Course*

In addition to contributing to the design of the course, we observed the *Creative Design and Engineering* class on 37 separate occasions averaging about two observations a week. We recorded video and audio almost every period that we observed. This consisted of two cameras and audio recorders that we rotated between small groups that we thought were engaging in more representation-heavy activities (e.g., looking up information about radiant heating on a website). We collected student notebook pages, photographs of artifacts, and design sketches. Additionally, we recorded and took notes in seven planning meetings with the teacher Ashley Freewoman. We conducted informal interviews with the students and Ashley as we observed, but did not conduct any formal interviews separate from class time.

Analyzing the Data

Overarching Approaches and Structures

As a qualitative ethnographic and design research study, our analysis was structured to explore, discover, and build theory around the uses of representations in making and making spaces. More specifically, the analysis of STEM literacies and the development of the framework was initially deductive, operationalized through a semi-closed coding scheme (Saldaña, 2015) that was developed using interview data and observations from the four distinct making spaces over three years in our project (Tucker-Raymond, Gravel, Kohberger, & Browne, 2017). This approach and analysis was designed to both shape the framework while also examining how our designs expanded or refined the working definitions of each literacy practice, contributing to both the empirical findings and the theoretical proposals of our study. After initial deductive analysis, we further inductively refined the themes we found in categories of literacy practices (e.g., identifying, organizing, and integrating information, or IOI). For instance, in IOI, the sources of information that people sought out was one theme, the social characteristic of IOI was another. Thus themes were created through a hybrid deductive and inductive approach and allowed us to focus on literacy practices while at the same time helping us identify emergent themes (Fereday & Muir-Cochrane, 2006).

Coding

Analysis of the data began with transcripts of interviews with experienced makers. To code, we used the initial version of our literacy framework, including the literacy practices culled from NGSS and CCSS. First cycle coding focused on giving the literacies some definition as well as identifying new codes relating to processes and activities that makers discussed (Schreier, 2012). We conducted multiple rounds of comparative analysis to refine the framework (Charmaz, 2014), discussing agreements and disagreements in weekly research meetings, and engaging in frequent "shop-talking" between meetings and onsite observations (Saldaña, 2015). First cycle coding was done primarily by hand on printed transcripts but was eventually ported to digital qualitative data analysis software. With a refined version of the coding scheme established from analysis of the experienced maker interviews, we conducted focused coding (Charmaz, 2014) of the observational and interview data from Artisan's Asylum and LMNPO. This yielded functional definitions of the literacies as well as categories like "values" that supported the development data-derived theoretical themes. These definitions were captured in thematic memos that included illustrative excerpts from different data sources, to reflect the results of the focused coding, and from which we were able to construct stories to share with participants and in publications about the study. Our final framework for literacy practices is in Chapter 1, Table 1.1.

Processes in Making Spaces

As described in Chapter 1, literacies are social practices enacted in particular contexts for particular purposes. Thus, our analytic approach also required means for characterizing the kinds of activities, tasks, and purposes that makers engaged with in their work (see Table BM1.2).

It was our intention to name making processes, while also taking care not to over-constrain the analysis. Carving individual's complex multifaceted practices into simpler, singular processes may mischaracterize makers' goals and motivations. Thus, we arrived at eight processes that we felt captured the work of making without over-specifying the observed episodes. The processes include the following: ideating, tinkering, designing or planning, fabricating, sharing, teaching, managing, and socializing. These unique, bounded processes served as the first unit of analysis, organizing our coding of interview and observation data to link literacies with particular processes, purposes, and goals of the makers. We labeled units of data with both a process and literacy practice because literacies are enacted within specific contexts for specific purposes. The coupling of literacies with processes anchored the analysis, and provided the infrastructure that allowed us to build theory over the life of the project.

These may not be the only processes that occur in making spaces. They also do not usually occur in isolation from one another, for instance many of the experienced makers we interviewed said that tinkering was part of their fabrication process. Likewise, socializing was integral to almost all of their time spent in youth making spaces and so it was integrated across the other processes. However, separating the different processes we have highlighted provided a frame for analysis when literacy practices might be most concentrated, complex, and relevant to learning STEM content and practices.

Member Checking

A critical component of our analytic approach was to present our findings back to the communities of makers that so graciously invited us into their spaces to study their work. Although validity and trustworthiness in our research come from the careful, intentional analytic processes we enacted (Guba, 1981), determining whether our findings were accurate to the participants' experiences was important to us. We approached this member checking in different ways: sharing of drafted manuscripts for feedback; hosting a community data session at Artisan's Asylum to share themes derived from our early coding work; involving participants as co-authors on publications for different audiences (e.g., research articles, conference papers); and having teachers reflect on their own observed practices using the STEM Literacies framework to explore how they saw these practices emerge in their own making. These sessions proved instrumental in helping to sharpen our findings and the ways we discuss them. Members of Artisan's Asylum, for example, initially balked at some of the ways we characterized their practices. In particular, around the idea of the Asylum operating with an "interrupting culture," where members communicated with each other frequently to seek and offer help and feedback. We presented

TABLE BM1.2 Processes in making spaces

Process	*Definition*	*Example*
Ideating	Captured in moments of inspiration, thinking, writing notes, doodling. Seeing, hearing, and feelings become the beginnings of constructions of meaning and ideas. People's experiences and frames of thinking, both disciplinary and everyday, serve as fodder for creativity (John-Steiner, 1997). People internalize these experiences and their multimodal representations, transform them, and externalize them as new ideas.	Yeah, [laughs] um but originally, I was like, "I want to fish. I want to be able to kayak or I want to be able to be on the water." Um and so I was like okay, I'll keep my ears open, and I'll start looking up videos of ways of how to build a boat—Um cheaply and so I was Googling like different types of boats and you can do them with plywood also and like carbon fiber or glass fiber, um but a week after I decided to make some sort of floating vessel for the Charles River, someone said that they were throwing away a kayak, right? So that was also a really big part. I was already sort of wanting to do that project.
Designing or Planning	"The iterative selection and arrangement of elements to form a whole by which people create artifacts, systems and tools intended to solve a range of problems, large and small" (Honey & Kanter, 2013, p. 3) Designing is goal-oriented problem solving driven by personal interest and the investigation of meaningful questions.	Um, a lot of times taking something that someone else has done and modifying it, that's both with software and with circuit design that's how I got started. I found somebody else had designed a keyboard and it didn't look anything like the keyboard I wanted but the basic circuits were right, and so I could, I think first started by trying to just change their design and then I just copied their, I mean this is really, there isn't a lot of intellectual property there, its key switches and diodes and rows and columns, those people have been doing for 40 years or 50 years, but I could see the insides of how somebody else had built it. Um and it's much much easier to get started on a project when all you have to do is do one little part of it to get to the next step rather than start from a blank canvas or at least for me.
Fabricating	To build or adapt objects by hand, for pleasure, economy, or as a way of understanding the world. Finding new lives and uses for old materials, creating new objects, or rearranging recognizable structures into new entities altogether (Honey & Kanter, 2013; Blikstein, 2013; Hatch, 2014).	Actually recently, I ended up, I just built an instrument a little before and I had a picture of it in my mind so I drew out a bunch of pieces on plywood and then I cut them all apart and that plan ended up being like a physical plan that I then assembled and then the sort of improvisatory process happened and then the sketching, building kind of conversation with the materials I had at hand sort of played out and then the instrument got finished a few days later.

Tinkering, Experimenting, Playing	Changing process in small or large ways and low-stakes testing of those ideas. Engaging with process and materials for exploration more than producing a product.	I experimented a lot. I tinker. Right, so. Um, you know if, you have the materials it doesn't matter if you mess up a pizza box or two, there's another one right there . . . learning through iteration, get to the, you know, get something in front of you and test it, stretch one of the parameters.
Sharing	Making one's work public after a finished project. Work is defined as both the process and product of a project. Sharing includes: telling, showing, (blogging; face-to-face show and tell, Instructables, Makerfaires) and selling (websites, Etsy shop, festivals). Can include sharing past experience of unfinished or continuing project.	But sometimes for example, I just started making some wooden spoons and I documented the process that goes into it and then put up a blog post. When the spoons are already available, so after the fact, but in details for anyone who's curious about how they were made. So that was on a blog on my website. Then there's obviously the sharing that happens, just like in the presence of other people and then I would say through emails, that's how I'll mostly update friends and family about stuff.
Teaching	Explicit instruction either in formal settings or informal settings where one person is recognized as an expert or leader that is facilitating learning experiences or sharing knowledge.	I found that I have made some sheets with the weaving laid out and I brought it back to one of the women that was in the group who had a particularly hard time and had—she started to color in, like just trace—color in where each—that ribbon shows. And then she is like oh, now I can see this part you are talking about like her brain still knew what the words were but now her eyes could follow it—and I'm like okay, this is it. I have to give everybody 10 minutes of coloring time before I even start to give them words because I am trying to use words for visual and they don't have the visual yet—and I am finding that you have to give people the developmental visual stuff first.
Managing	Includes references to organizing making activities or spaces in which people engage in making activities. It may include talk about managing employees, interns, or apprentices for one's own project as well as in general. Identifying and repairing broken tools.	Even just talking to people, we were talking to a bunch of people about like how do we design the linkages. How, like how are we going to communicate. 'Cause we had to make hundreds and hundreds and hundreds of pieces for it. And we were coordinating a group of people who had never done a lot of this type of stuff before.
Socializing	Conversation or interaction not in line with our activities or literacy practices. Talking in friendly ways about topics not related to making.	Trovas and GMH are debating whether Wendy's or Tasty Burger has better food.

excerpts from observations and interviews to expound on our analysis. Seeing their own words and work presented back to them, through the lens of the themes our analysis produced, resulted in affirmational, yet also illuminating, conversations. We learned that "interrupting" was, in fact, a central practice of the makers at Artisan's, but that for some this feature of the space was unwelcomed, and thus those individuals ceased their participation in the community. In sum, the varying modes of member checking helped to strengthen the framework that we present in this book, by both sharpening the ways we define and describe the literacies, and in adding depth and complexity to the cases we present.

Designed Interventions

Our goals for this study were not only to observe the literacies that makers enact, but also to explore how we might be able to support the development of STEM literacies within making. As such, we employed design-research approaches in developing interventions that we felt would support learners. We designed interventions in the *Creative Design and Engineering* class collaboratively with the cooperating teacher, Ashley Freewoman. We also worked with Ashley and our research team to design the teacher workshops and study group.

These interventions allowed us to more pointedly examine particular aspects of literacies by designing for specific kinds of activities. For example, we designed protocols for youth makers to develop "Tutorials" or "How-tos" for using particular tools or procedures in making (see Appendix B). We observed both adult and youth makers turn to YouTube with great frequency to seek help for whatever challenge they faced. We wondered whether asking youth to author these kinds of videos would give us insights into how they used literacies to communicate what they were learning about making. In one tutorial, titled "THE BEST SCREWING TUTORIAL!!!!!" two ninth grade makers explained how to properly put a screw into a piece of wood. In making the video, they encountered aspects of their own understanding that they wanted clarification about: how does it work to tighten the bit using the chuck? Why drill a hole first? The intervention—making video tutorials—served as a structure to support youth in reflecting on and refining their understandings. This is an example of how interventions were used to better explore how literacies may be practiced in making. In the example, authoring the video was a form of reflection, which relates closely to how we think about sharing designs in engineering, or results in science.

Case Studies

In the initial conceptualization of the work we wanted to create empirically-based case studies of making spaces. At the same time, we wanted those case studies to describe all the ways members of the spaces used literacy practices. This proved to be an easily overwhelming task. We decided that we would use each space to highlight one of the literacy practices from our framework. We had four literacy practices in our framework and we had sufficient data from four spaces that we observed. It was a match. However, we also wanted to be able to analyze literacy practices across spaces and so we created some of our stories about literacy practices to draw from multiple spaces.

We decided on which literacy practice to pair with which space depending on what kinds of literacy practices were emphasized in our data. Moreover, cases were developed based on the forms of data we collected at the different sites; AA and LMNPO were exploratory, intended to map the space of literacies in particular spaces. Data from these spaces were mainly composed of interviews and observations. At SETC and at Nedlam's Workshop, we were able to engage in more intensive observational work and we used video and audio recordings to capture particular moments of making, to drill down into the specifics of how the literacy practices unfolded moment by moment, in making. Each case was built around the emergent themes for the assigned literacy practice. For instance, the Artisan's Asylum case study, while about the practice of communication explores the themes of distributed expertise, interdisciplinarity, and collaboration. It does so by describing how the space and social practice are co-constitutive.

However, by now the reader will have noticed that the cases are not solely of a space or only about a literacy practice as we outlined in our initial table. Rather, they are about how that practice manifested in the context of the space's values, relationships between people, and material conditions. The cases are descriptions of how literacies help to understand forms of work including ways of communicating, and consuming and producing representations, in social context for goal-directed purposes.

References

Blikstein, P. (2013). Digital fabrication and "making" in education: The democratization of invention. *FabLabs: Of Machines, Makers and Inventors*, *4*, 1–21.

Charmaz, K. (2014). *Constructing grounded theory*. Thousand Oaks, CA: Sage Publishing.

Emerson, R. M., Fretz, R. I., & Shaw, L. L. (2011). *Writing ethnographic fieldnotes*. Chicago, IL: University of Chicago Press.

Fereday, J., & Muir-Cochrane, E. (2006). Demonstrating rigor using thematic analysis: A hybrid approach of inductive and deductive coding and theme development. *International Journal of Qualitative Methods*, *5*, 1–11.

Geertz, C. (1973). Thick description. In *The interpretation of cultures* (pp. 3–30). New York, NY: Basic Books.

Gravel, B. E., Tucker-Raymond, E., Kohberger, K., & Browne, K. (2018). Navigating worlds of information: STEM literacy practices of experienced makers. *International Journal of Technology and Design Education* *28*(4), 921–938.

Guba, E. G. (1981). Criteria for assessing the trustworthiness of naturalistic inquiries. *ECTJ, 29*(2), 75–91. https://doi.org/10.1007/BF02766777

Hatch, M. (2014). *The maker movement manifesto: Rules for innovation in the new world of crafters, hackers, and tinkerers*. New York, NY: McGraw Hill.

Heath, S. B., & Street, B. V. (2008). *On ethnography: Approaches to language and literacy research. Language & Literacy (NCRLL)*. New York, NY: Teachers College Press.

Honey, M., & Kanter, D. (2013). *Design, make, play: Growing the next generation of STEM innovators*. New York, NY: Routledge.

John-Steiner, V. (1997). *Notebooks of the mind: Explorations of thinking*. Oxford: Oxford University Press.

Jordan, B., & Henderson, A. (1995). Interaction analysis: Foundations and practice. *The Journal of the Learning Sciences, 4*(1), 39–103.

Saldaña, J. (2015). *The coding manual for qualitative researchers*. London: Sage Publishing.

Schreier, M. (2012). *Qualitative content analysis in practice*. London: Sage Publications.

Sheridan, K., Halverson, E. R., Litts, B., Brahms, L., Jacobs-Priebe, L., & Owens, T. (2014). Learning in the making: A comparative case study of three makerspaces. *Harvard Educational Review, 84*(4), 505–531.

Tucker-Raymond, E., Gravel, B. E., Kohberger, K., & Browne, K. (2017). Source code and a screwdriver: STEM literacy practices in fabricating activities among experienced adult makers. *Journal of Adolescent & Adult Literacy, 60*(6), 617–627.

APPENDIX B

Sharing Genre Scaffolds

Guidelines for Crowdsourcing Proposal

Crowdsourcing is a way for people to fund projects that they cannot fund on their own. In trying to get funding for what you are making, first research what others are doing and the ways in which they have been successful.

Research

Read This Article and Answer the Following Questions

www.entrepreneur.com/article/228534

- Is this web article credible?
- Why or why not?
- Why might it be biased?
- What are the different types of crowdfunding platforms?
- What are the pros and cons for each of the websites in the *reward* section?

Go to the Websites Listed Below and Explore a Few of the Successful Projects on Each

Kickstarter: www.kickstarter.com
Indiegogo: www.indiegogo.com
Rockethub: http://rockethub.com/
Peerbackers: http://peerbackers.com/
iFund Women: https://ifundwomen.com/

Write down four elements of presentations that these successful projects seem to share. Do you think you can incorporate any of these elements in your presentation? How so?

Making a Project Page

Have you made your prototype? Good. Time to start planning for your crowdfunding page.

Telling Your Story

A video is strongly encouraged but not required. More than 80% of projects on Kickstarter have videos, and those that don't have a much lower success rate. Here are questions to help you tell your story, and make your video.

1. Who are you?

    Introduce yourself and your team and any similar work you've done.

2. What are you planning to make?

    The more details the better. Include photographs and descriptions of any sketches, samples, prototypes you've made.

3. Where did this project come from?

    Tell people how you got the idea, and how much you've accomplished so far. Sharing the project's history helps others understand the kind of work you do and how you go about it.

4. Why is it important to make and what community issue will it address? Who will use this project?

    Tell people why you are passionate about your project and why you are committed to making it.

5. What technologies will you use?
6. How much will this product cost the consumer?
7. Choose an image that shows off your project. This will go on the top of your project page.
8. Make a video about your project. (see guidelines for making a video)

    Pick a place in your narrative to show off your video.

(Optional) Competition: Your crowdsourced project will be voted on by members of your school. The winning project will receive extra funding to put your idea into production.

Guidelines for Creating an Instructables Post (www.instructables.com)

1. Document everything on a blog, in a doc, or in your notebook. Take pictures. Write down all your steps and your difficulties.
2. Take a photograph or make a short video (less than 20 seconds) of your final project so that others can see what they will be making. Think about the angle and distance that puts your project in the best light.
3. If you are taking a photograph use indirect sunlight without a flash when possible. If not possible, you can edit it later. Make sure the area where you are taking the photo is clean so that your project is the focus. Use a tripod if you can.
4. You will want to include photos for each stage (but not too many). Think about what photos you want to use before you get started. Look through your photos and select a few for each step. You can always change them.
5. You can edit photos for brightness, contrast, saturation and you can crop them in a photo editor. You can edit photos in the photo editing program on your computer. GIMP is a free one: www.gimp.org/
6. You can also put photos together in a collage to show small or related steps. This is an Instructables post on how to edit: www.instructables.com/id/basic-photo-editing/.
7. Make a profile on Instructables. You do not have to add a photo of yourself, but you should say a little bit about who you are and the kinds of things you are interested in making.
8. You will use the Instructables editor to upload your project directions. Instructables provides a tutorial on how to use the Instructables editor: www.instructables.com/id/How-to-Use-the-Instructables-Editor/?ALLSTEPS
9. On the next page you can outline your Instructables before you get started.

Instructables Outline

Use as much space as you need. If you do not have photos, that is okay. What might you put there next time?

Introduction Page:
Short description of project. Why you made it.
Any needed safety rules

Photo/video to include:

Step 1:
Title:
Description of STEP

List of photos/video to include

Step 2:
Title:
Description of STEP

List of photos/video to include

Step 3:
Title:
Description of STEP

List of photos/video to include

Step 4:
Title:
Description of STEP

List of photos/video to include

Step 5:
Title:
Description of STEP

(Continue with as many steps as needed)

Guidelines for Creating YouTube "How-To" Videos

You are learning to do things in the making space. Like how to use a drill, how to use electronics, how to load an image for the laser cutter, etc.

YouTube "How-To" videos can be really helpful for other people learning to use these tools, or how to do similar processes, in the making space.

Making a "How-To" video can be as complicated, or as simple, as you would like.

1. Spend some time looking at How-To videos on YouTube. Maybe look up tools that you see in the making space, like scroll saws or hand drills.
2. Once you decide on what tool or skill you will make the video about, write a short script. You do not have to write down every word you will say, but you there are some things you should include:

    a. Who are you? And why are you making this video?
    b. What tools, process, or technique are you showing?
    c. How did you learn this?
    d. What are the materials you will need?
    e. Any safety considerations before you start?

f. Create the steps of the demonstration.
g. For example, for drilling a screw into wood:

    1. Secure the wood.
    2. Put a drill bit slightly smaller than the screw diameter in the drill.
    3. Drill a hole where you want the screw to go, going about ¾ of the depth of the screw.
    4. Load a screw bit into the drill.
    5. Screw the screw into the hole using the drill.

3. When you are ready to film, enlist a friend to help. That person could hold the camera. You could also consider using a head-mounted camera, like a GoPro or putting a camera on a tripod. You would have to be the narrator, talking aloud as you did the technique or showed how to use the tool.
4. Once you have made the video, have someone look at it and give you feedback. Videos are pretty quick to make, so a second draft could help improve the quality of the video.
5. When ready, upload to YouTube! Think about the title, the category, and keywords you might include. You want people to find this video when they look, and you want it to be helpful.

APPENDIX C

Meshwork Mapping Protocol

Overview of Protocol: This protocol is a suggestion of how to structure an activity that encourages youth, particularly youth in school making spaces, to begin constructing maps of their meshworks. Chapter 4 illustrates how meshworks are a useful idea for supporting the cultivation and maintenance of dynamic networks of expertise that support the flow of information. Conceptualizing the space, and those in it, as existing in this meshwork values the distributed expertise, multimodality, and transdisciplinarity in making spaces. The activity also builds an infrastructure to support learning.

In this activity, youth work to explicitly name the people and places they go to for help and information. They talk about this in terms of individuals, but also in terms of categories and domains. They begin to construct maps that look much like networks and that highlight the connections they have across domains and settings that can support them in their searches for information. They then contextualize this within the making space, and the domains and areas of expertise that are useful to their projects. The representations they create are static, but we encourage this activity to be repeated, where the maps are revised and refined as new points of connection, new information, and new expertise are developed. We imagine this activity could be extended, for example, perhaps the entire community constructs a meshwork on a large wall, where all the points of connection are represented. New connections, new sources of information, and even gaps in expertise are added to the wall. Ultimately, the goals are to surface the existence of these meshworks, making them visible and accessible to support youth learning STEM in making spaces.

This activity should be implemented more than once, over extended periods of time.

Materials: Blank 11x17 paper, blank poster paper, markers, pens, pencils, sticky notes

Introduction

Facilitator reads a summary of the activity:

We want to get you to think about **who** and **where** you go for help and information for different types of problems in your life, at school, and at work.

In a minute, we'll ask you to put your name in the middle of the paper. We will ask you to add the names of other people and places you go to for information, for example, your Church community or the Youth Center. These should be places connected to you, people you talk with often when you have questions or need information. We'll be making something kind of like a concept map, if you are familiar with those. (Facilitator show an example)

But first, we'd like to talk about questions you have when you are at **school.** For instance, when you need to know about homework assignments.

Getting Started—*Generating a List*

Hand out large pieces of paper—11×17—with pencils, pens, markers.

Facilitator

- Before writing, we want you to think a little about some questions. Then we'll get to writing things down.
  a. Who do you go to for help or to find information?
  b. Where do you go for that help?
  c. Think about that for a minute. If you want to write something down, that's fine.
  d. Now think about whether you go to some people, places, or sources more than others? Who? Why?
- Ok, now, make a list of the names of the people/place/sources that you go to on the sheet of paper we gave you.
  a. List the names, places, and other things that came to mind in a list on that paper.
  b. Turn to the people at your table, and talk about who is on that list, and why you put them there.

Whole-Group Discussion—*Ideas, People*

Facilitator

- What *kinds of information* do you need to know about at school? [Encourage participants to share what they have written or talked about]
- Who do you go to for *which kinds* of information? For example, do you ask different things of teachers than you do a principal?
- Do people ever come to you with questions? About what?

Creating Meshwork Representations—*From Lists to Maps*

Facilitator

Now we'll transform this list into a map—or a web—of the people, places, and information you talked about. On the piece of paper in front of you, put your name in the middle and circle it. Now, make a map of the people you go to with questions, or to find information. Include stuff about the kinds of questions you ask, the topics you discuss, or the advice you seek and share. You can organize this however you would like. Take a few minutes to do this.

Once done, turn and talk to the other people around you. Compare and contrast your maps: Do you have similar categories of people—friends, teachers, parents—or similar categories of topics—homework, jobs, college? If not, what's different? Are the things you put down only about school? Or other aspects of life?

Group Discussion—*Sharing Maps: From Individuals to Domains of Expertise*

Facilitator

Would a volunteer share what is on your map? [Ask at least 2 people to share their maps. As they share, facilitator writes down some of the categories of information (e.g., homework, jobs) that you hear on sticky notes, one sticky note at a time. Place these on poster board.] As you see, I've been keeping track of the kinds of information and the kinds of questions you ask other people in your lives. Please look at your own maps and write down what you have there—not the people, but the kinds of information, or categories of help that you seek. Please only put one idea on a sticky note. Include things that people ask *you* about. When you are done, place the sticky notes on this poster board.

Sorting Activity: *Finding Shared Domains*

Facilitator encourages the group to begin sorting and arranging their sticky notes according to categories or similar kinds of topics/domains. Once sorted and organized, write the names of people in the network next to these categories, drawing from each of the individual meshwork maps. Perhaps rename the categories to reflect some overarching themes. This generates a shared map of expertise, people, resources, etc.

Reforming Maps: *Moving from School to Making Space*

NOTE: This should be done after participants have begun working on a project, where they are facing particular needs relative to the dimensions and constraints of the problem.

Facilitator

[Facilitator hands out the individual maps generated previously]

We did great work last time mapping the information we seek in school. We generated some categories of information, or domains of knowledge and expertise related to life in school. Now, when you are at [making space] or need information related to your work here, where do you go? To whom to you give ideas or information, and where do you share information?

- [Making space] is in your school—are there people on the school list that are also on the list of people you might go to for help in [making space]?
- Who else? Now that you are working on a project, who comes to mind as people who know things that you might want to ask questions of?
- Looking at your existing map, either add to the existing one, or draw a new one and arrange people in the ways you want so that you can capture new ways of thinking about information and advice you need, specific to [making space].

Information Flow: *Thinking About Direction, and Flow of Information*

Facilitator

Looking at your maps, think about the directions that information flows. If you ask people questions, draw an arrow from you to that person on the map. If they ask you questions, make it a double arrow. Use the arrows to capture the ways information gets exchanged. Double arrows mean that information goes both ways.

Now you have both of these networks of information about School and [Making Space]. Let's look at the shared map, do we have people to add? Categories to add? What's not captured here?

Extensions

- *Possible sites of information*: Create a next layer of other possible sites of information: including public institutions and private businesses (e.g., library, theater group, hardware store, twitter). Within the context of their making, ask participants to think expansively about where they might find useful information. For example, churches, community centers, and government agencies.
- *Addressing gaps in meshworks*: "Where are gaps in the maps? What are things you think you might need help figuring out, or things you think you need to know, but that aren't represented here?" Naming domains, and then considering new areas (e.g., designing simple circuits), can help think about people within the meshwork who are not captured on the maps (e.g., a student knows an electrician, but hadn't thought about that as a relevant skill in making before). "Are any other sources of information that *you know* about that you could help others with?"

Additional Follow Up Questions About the Activity

- What did you think about this activity?
- Do you think you learned something about yourself?
- Do you think you learned or re-learned something about anyone else?
- Do you think making this map will help you in the future?
- What do you think about your role as a source or contributor of information? How would you want people to get in touch with you?
- Would you be interested in creating a more permanent map for [making space]? So that if people had a question they would be able to go to that place/person? What do you think that would look like? What form would you want it to take (physical/online)? How would you update it?

APPENDIX D

STEMLiMS Object-Oriented Interview Protocol

Hi, [NAME]. Thank you for agreeing to talk to me today. I really appreciate it. So, my colleagues and I are doing some research to learn more about people who like to build, create, or hack different kinds of things or artifacts in the world.

We are doing some interviews with a select group of people from a variety of fields, including you, to understand your involvement in making, building, or hacking things, how you have learned the things you need to learn, and most of all, how you use representations such as writing, drawing, and modeling as part of your processes. Okay?

In this interview I'm going to probably use the word "making" a lot because it's the popular discourse, the magazine, the faires, and the language that the government uses to identify this kind of activity or work. You may use a different word or words for what you do, or a group that you identify with, and over the course of the interview I'd love to hear your perspective on what that is. Okay? Do you have any questions?

[Second and subsequent questions within a given number (2.) are asked one at a time. Do not ask more than one question at a time.]

1. First, you know that this is a study about what is popularly called, "making." The first question is pretty open. I would just like you tell me about yourself as someone who makes things.
2. (If necessary) How did you get into (making/creating/hacking) things? How long have you been doing it? Why do you do make things? (e.g., work/play relationships)
3. (If necessary) Where do you make things?
4. (If necessary) Are you involved in science, technology, engineering or math in your non-making times? Do see that as related to your making?

5. When you want to learn how to do something new that is related to your making, how do you go about it?
6. Can you tell me some of the things you made and think about one project in particular that we can talk about in more depth? (Why did you make it? Where did you make it? With whom?) In thinking about that project:

   a. What materials did you use (if relevant)?
   b. Did you pick up any new skills? Or refine and develop any old ones? Were there different skills you learn at different times in the project?
   c. Did you learn anything new about your work?
   d. Did you work with other people on this project? If you worked with other people, what sorts of skills do they have? What kind of people did you seek out?
   e. If you needed to learn something to do the project, how did you find out what you needed to know? Do you think of the knowledge that you needed as existing in any particular area of expertise or knowledge? Do you think of those sources as from multiple or combined areas of expertise?
   f. (If necessary) Did you encounter any challenges while making this___________? How did you deal with or solve them? (follow up: did you use other people? other forms of information? trial and error?)

      i. If you got more than one opinion about how to solve this problem, how did you determine the best advice? How do you sift through competing opinions to make a choice for yourself?

7. We are thinking about how you use different kinds of *representations* to make this object. So, representations would be words, both written and oral; sketches; analog and electronic images that might be moving or static that you make or view; models of all kinds; schematics; numbers that you might find in equations or measurements. Also, we're thinking of the making process as consisting of three or four stages: 1) inspiration and/or planning, 2) playing or tinkering, 3) making, and 4) sharing . . . What is your process like? Does it resemble what we have sort of outlined here?

   a. So what is your planning process for a new object? Can you give me an example of how you use representations?
   b. What's your fabricating/making/hacking process? Can you give me an example of how you use representations?
   c. How do you share your work? Can you give me an example of how you use representations?

8. Finally, is there anything else you've thought of while we are talking that you think is relevant?

Thanks again [NAME]. We really appreciate your time and your expertise. Is it alright if we maybe talk to you in the future with some follow up questions?

INDEX

Page numbers in *italics* refer to figures, page numbers in **bold** refer to tables.

CPSIA information can be obtained
at www.ICGtesting.com
Printed in the USA
LVHW031508161121
703494LV00010B/864

9 780815 367741